12/30/93

D0072887

Modeling Complex Phenomena

Woodward Conference
San Jose State University, San Jose, California

Wave Phenomena: Theoretical, Computational and Practical Aspects
Lui Lam and Hedley C. Morris, Editors

Nonlinear Structures in Physical Systems: Pattern Formation,
Chaos and Waves
Lui Lam and Hedley C. Morris, Editors

Modeling Complex Phenomena
Lui Lam and Vladimir Naroditsky, Editors

Lui Lam   Vladimir Naroditsky

Editors

Modeling Complex Phenomena

Proceedings of the Third Woodward Conference
San Jose State University
April 12–13, 1991

With 116 Illustrations

Springer-Verlag

New York Berlin Heidelberg London Paris
Tokyo Hong Kong Barcelona Budapest

Lui Lam
Department of Physics
San Jose State University
San Jose, CA 95192, USA

Vladimir Naroditsky
Department of Mathematics and
   Computer Science
San Jose State University
San Jose, CA 95192, USA

Library of Congress Cataloging-in-Publication Data
Woodward Conference (3rd : 1991 : San Jose State University)
    Modeling complex phenomena : proceedings of the Third Woodward
Conference, San Jose State University. April 12–13, 1991 / [edited
by] Lui Lam, Vladimir Naroditsky.
       p.   cm.
    Includes bibliographical references and index.
    ISBN 0-387-97821-6. — ISBN 3-540-97821-6
    1. Chaotic behavior in systems — Congresses.   2. Computational
complexity — Congresses.   3. Mathematical physics — Congresses.
I. Lam, Lui.   II. Naroditsky, Vladimir.   III. Title.
Q172.5.C45W66   1991
003'.7 — dc20   92-4667

Printed on acid-free paper.

Production managed by Henry Krell; manufacturing supervised by Robert Paella.
Camera-ready copy prepared by the editors.
Printed and bound by Edwards Brothers, Inc., Ann Arbor, MI.
Printed in the United States of America.

9 8 7 6 5 4 3 2 1

ISBN 0-387-97821-6 Springer-Verlag New York Berlin Heidelberg
ISBN 3-540-97821-6 Springer-Verlag Berlin Heidelberg New York

Preface

Once upon a time, science was not divided into disciplines as we know it today. There was no distinction between so-called social and natural sciences, not to mention the fragmentation of the latter into physics, chemistry, biology, geology, etc. According to legend, the scientists those days would do their research in whatever environment they happened to find comfortable, which more often than not was in bathtubs or giant hot tubs — remember Archimedes!

Then, somehow, these days we find ourselves compartmentalized into different departments in our universities, or divisions in our research institutes. (We suspect, for one thing, that is to ensure that we will get our paychecks delivered on time at the end of each month.) Anyway, as anyone who has worked in the *real* world knows: when one is confronted with a completely new problem or phenomenon, it is usually impossible to neatly assign the problem to physics, chemistry, or, for that matter, computer science. One needs to recall and fuse together the knowledge one learned before and, if that alone is insufficient, to consult experts in other areas. This points to the shortcomings of the compartmentalization of knowledge in our educational systems.

In recent years, something has changed. Under the banner of *Complex Systems*, some brave souls are not afraid to tackle problems that are considered intractable by others, and dare to venture out of their trained disciplines or departments to which they are attached. A psychological barrier is broken, and the good old days of one unified science could be back!

While a clear definition remains elusive, Complex Systems research usually signifies a synthetic approach in the investigation of systems consisting of a large number of interacting, simple components, irrespective of their origins. In fact, common features are found and common methods are employed in the study of many social and natural systems. Many of the best brains in the world are now engaged in this exciting, new frontier of science. To find out why and what they have been doing, we invited a group of leading experts here at San Jose for two days on April 12-13, 1991, at the Third Woodward Conference on *Modeling Complex Phenomena*. The use of the words "complex phenomena" instead of "complex systems" emphasizes the fact that we are interested in real problems from the real world.

In their talks, the invited lecturers were asked to give a long introduction to the background materials, summarize the major findings and present new research results. The topics covered include physical computation, nonlinear forecasting, machine learning, neural networks, fuzzy logic, chaotic dynamics and economic models, artificial life, earthquake modeling, and complex patterns. The papers in the proceedings here include both invited lectures and posters. They are loosely divided into six parts according to the phenomena involved, rather than the research methods used. Chaos and self-organized criticality are the two recurrent themes common to some of the papers.

Part I consists of three papers on paradigms, complexity, and learning. Part II is on nonlinear forecasting and the arms race. Economic systems make up Part III, while earthquakes and sandpiles are grouped in Part IV. Part V is about computations in fluids and crystal growths. Finally, various problems in complex patterns are contained in Part VI. For a more detailed description, the reader is referred to the first paragraph of each article which serves as the abstract of the whole article.

The conference was sponsored by the Center for Applied Mathematics and Computer Science, the Department of Physics and the Society of Archimedes, all of San Jose State University, and was funded by the Woodward Bequest. The Advisory Board consisted of David Campbell (Los Alamos), Jim Crutchfield (Berkeley), Chris Langton (Los Alamos), Horazio Mendez (IBM San Jose), Roald Sagdeev (College Park), and Patrick Suppes (Stanford). Lui Lam and Vladimir Naroditsky served as cochairs of the conference; the Organizing Committee consisted of Alejandro Garcia, Valery Kanevsky, Lui Lam, Igor Malyshev and Vladimir Naroditsky.

We would like to thank all the people involved, especially to the invited speakers — Michele Boldrin, Martin Casdagli, Jim Crutchfield, Richard Durbin, Alejandro Garcia, David Haussler, Alfred Hübler, Valery Kanevsky, Bill Langlois, Chris Langton, Gottfried Meyer-Kress, Patrick Suppes, Chao Tang, and Lotfi Zadeh — for their skillful presentations. We are particularly indebted to Jim Crutchfield who helped to shape the conference at the beginning and persuade others to write up their talks at the end. We are grateful to Arlene Okerlund, our Academic Vice President, for delivering the welcoming remarks which are included in these proceedings; to Alan Ling, Veril Phillips and Donald Strandburg for their enthusiastic support.

Well, like many of the participants, we did enjoy very much and learned a lot at the conference; we hope the readers will do the same by reading these proceedings. With more progress and a little bit of luck, maybe one day we will again be able to do science together in the swimming pools, assuming, of course, that water-proof laptops are available.

San Jose
December, 1991

Lui Lam
Vladimir Naroditsky

Contents

A Complex Phenomenon: The Human Mind

A.N. Okerlund

On behalf of San José State University, I am honored to welcome you to the Third Woodward Conference, sponsored by the Center for Applied Mathematics and Computer Science, the Department of Physics and the Society of Archimedes.

Your presence here is the result of one of those fortuitous events that sometimes take us quite by surprise. There is a very simple lesson in this conference that I ask you to keep foremost in your mind throughout your two days of sophisticated, highly complex discussions of computer-generated models. I ask you to remember that your considerations of such a topic as "Chaotic Dynamics and General Equilibrium Theory" are made possible by an one-time student at San José State. This conference is funded in memory of Henry T. Woodward, who received a Masters degree in Mathematics from this University in 1958. Mr. Woodward went on to work at NASA Ames Research Center with the Planetary Atmospheres Group until his death in 1984. Upon the subsequent death of his mother, Marie Woodward, in 1986, San José State University received a bequest in memory of her son, the bequest that has endowed this conference.

In short, your discussions of "Modeling Complex Phenomena" literally have been made possible by the positive educational experience of a single student. That point is especially intriguing in the context of today's national educational controversies about the research agenda of universities dominating their teaching missions. At San José State, we have always emphasized the importance of good teaching, and we insist that good teaching is the *sine qua non* for tenure and promotion. One never knows, after all, which student may go on to win the Nobel prize, develop a model of earthquakes, or work at NASA and bequest half a million dollars to his or her *alma mater*.

That good teaching contributes to research is confirmed by another example I would like

A. Okerlund: Academic Vice President, San José State University, San Jose, CA 95192

to relate: one of our undergraduate students who has just received an ARCS scholarship for her achievements in mathematics first arrived at SJSU as an English major. Because our General Education requirements mandate that all students must take a math course, Susan Hansen (please note the female gender) did exactly that. She found the required math course to be so interesting that she took another. And another. Ultimately, she changed her major to mathematics, and she will graduate with a B. A. this semester. She has applied to graduate school, where her objective is a Ph.D. in Applied Mathematics with a focus in "Dynamical Systems." By the time the Woodward Conference celebrates the end of its first decade, I fully expect that Susan will be on the program reading a paper about time evolution of complex phenomena, a paper every bit as complicated as the ones you will be hearing today and tomorrow.

In welcoming you to SJSU, therefore, I would like to emphasize one positive point which contradicts the negative press that universities are receiving these days: the educational system *is* working. Good teaching *can and does* occur. Good teaching *can and does* promote great research--sometimes in curious and totally unpredictable ways. Thus, as you disseminate the results of your research under the auspices of the Woodward Conference, I hope you will pause a moment to reflect upon the genesis of this event. And when you return to your classrooms, I hope that you will remember the crucial significance of your interactions with each of your students, every moment of which becomes part of a new and extraordinarily complex phenomenon: the development of a human mind.

Part I
Paradigms, Complexity, and Learning

Modeling and Control of Complex Systems: Paradigms and Applications

A. Hübler

In many cases, the dynamics of high dimensional nonlinear systems can be estimated from a low dimensional model. Nearly all variables are slaved by a few order parameters. If the complex system is perturbed by an external force in order to control it or to investigate it with a spectroscopic method, slaved variables can be stimulated and the prediction of the response from the low dimensional model may be impossible. We show that it is generally possible to predict the response of the complex system and to control it, if the external forces are resonant perturbations of the low dimensional model. We present this issue in terms of a few paradigms including the principle of the dynamical key, the principle of optimal interaction and the principle of matching in the framework of other paradigms in complex systems research. We discuss open problems as well as possible industrial applications.

1. INTRODUCTION

Complex systems research has made a tremendous progress during the last few years. Concepts like synergetics and slaving[1], neural nets[2], fuzzy logics[3], genetic algorithms[4], artificial life[5], deterministic chaos[6], fractal geometries[7], solitons[8], catastrophes[9], and self organized criticality[10] have been used to model typical phenomena of complex systems, i.e. high dimensional evolving systems. The progress was triggered by advances in computer technology, since digital computers are good chaos generators. This progress is expected to have a strong impact on industrial technology in the near future since the chaotic signals produced by computers are ideal tools to interact with real complex systems[11, 12] and to control them[13, 14, 15, 16]. That results from the fact that nonlinear systems react most sensitively if the period of the perturbation fits to their actual internal rhythm[17, 18, 19]. This is the principle of the dynamical key. We present appli-

A. Hübler: Center for Complex Systems Research, Department of Physics, Beckman Institute, University of Illinois, Urbana, IL 61801

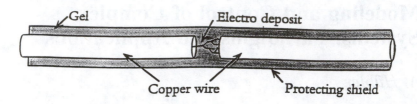

Figure 1: *A self repairing electrical cable out of a copper wire, surrounded by gel with copper ions and a protecting shield. When the copper wire is broken, and a voltage is applied, the wire will reconnect due to electrodeposition. The deposit has under usual experimental conditions a complicated, dendritic structure. The dynamics of the dendritic growth is complex.*

cations of this principle which show that dynamical properties of some complicated experimental systems can be determined orders of magnitude more accurately if sine generators or sources for monochromatic electromagnetic waves are replaced by chaos generators. In addition, we show that it is possible to get sharp diffraction peaks from disordered materials if the incoming wave is chaotic. This makes it possible to determine the structure of disordered materials accurately. A better understanding of disordered materials could make it easier to operate industrial processes in strong nonequilibrium, since physical systems tend to be disordered in strong nonequilibrium. A large flow of heat, a large current or a large flow of any other substance penetrating a physical system usually causes disorder, for example amorphous or fractal structures or turbulent dynamics. Some of those physical systems in strong nonequilibrium have the capability to repair minor damages themselves. They are regenerative(see Fig. 1). The investigation of regenerative materials may trigger the production of self repairing electronic equipment. The principle of minimum resistance and the principle of optimal interaction explain properties of regenerative materials.

Those principles and, in particular, the various control techniques for chaotic systems may be applicable in social and economic systems too. Computers are speeding up all business interactions, whereas management techniques are still based on personal experience and judgement. In order to match the speed determined by computer based business, management has to be computer assisted, too. Since economic systems are complex nonequilibrium systems[20] and time scale arguments are most important in physical systems as well as in economic systems, the principle of the dynamical key in combination with the existing knowledge in complex systems research should provide the right tool for an efficient control of turbulence as well as computer assisted management.

A dramatic change in experimental equipment and materials paired with a broad applicability of the new concepts could produce a substantial technological breakthrough. Many systems in everyday life and in industry exhibit chaos, turbulence, and catastrophes. Chaos, turbulence and catastrophes are frightening people because they are generally assumed to be uncontrollable and unpredictable. People try to avoid them or to limit the "negative" influences of chaos, turbulence and catastrophes to a minimum. The method of dynamical control enables us now to handle some of the most sophisticated, unpredictable systems, including some systems exhibiting chaos, turbulence and catastrophes and to use them for scientific and industrial purposes. Therefore it is necessary to replace the labels "negative" and "unacceptable" for features, like chaos, turbulence, catastrophes and slaving, by a careful scientific classification. The goal of this article is to present some new theorems, paradigms and conjectures in the framework of existing methods. In the context of some physical examples we present a mathematical background for these paradigms and colloquial interpretations. We hope that this facilitates the transfer to other disciplines. In addition, we emphasize possible industrial applications. The next few sections provide an overview of paradigms, theorems and conjectures on (i) defining and producing complex systems, (ii) modeling and control of those, and (iii) general rules extracted from those models. The last section gives a summary and a perspective on future developments.

2. MAKING SYSTEMS COMPLEX: THE NONEQUILIBRIUM COMPLEXITY PARADIGM

A key strategy in complex systems research is to investigate physical systems where the amount of complexity can be adjusted through the degree of nonequilibrium. There are two different classes of nonequilibria: internal and external nonequilibrium. A physical system is in internal nonequilibrium if it is at a transient or metastable state. Examples are glasses, amorphous solids and snow. A system is in external nonequilibrium if it is penetrated by a given constant flux, a sinusoidal flux or an aperiodic flux. Examples are structures of stationary liquid-melt interfaces where the complexity is adjusted through the flow of heat through the interface[21], fluid flows behind hindrances where the complexity is adjusted through the Reynolds number, or macromolecules under the influence of a chaotic field, where the complexity is adjusted through the type of chaos[19]. Table 1 shows a classification of external nonequilibria with some examples of the corresponding complex systems.

Table 1: *Examples for low dimensional and high dimensional physical systems in an external nonequilibrium.*

Categories	low dim. system	high dim system
constant flux	laminar flow of a liquid	dendritic crystallization oscillating chemical reactions turbulence behind a hindrance
periodic flux	chaos in nonlinear mechanical systems quantum chaos	induced turbulence
chaotic flux	nonlinear resonances controlled chaos induced chaos	controlled macromolecules controlled turbulence induced turbulence

Generally it is assumed that *the complexity of the state of an experimental system increases when the nonequilibrium increases* (<u>nonequilibrium complexity paradigm</u>). However, up to now no measures for complexity or nonequilibrium are known, which would make it possible to put this paradigm in a quantitative relation.

It is a striking fact that most industrial products are processed in strong external nonequilibrium or operate due to external nonequilibrium whereas most physical theory does not apply for external nonequilibrium and for example most experiments in condensed matter physics are done in the absence of strong external nonequilibrium. Therefore in many cases the intuition of an experienced engineer combined with a trial and error search strategy is the only source for improvement of industrial processes in particular if the industrial process does not operate at a stationary state. Let us assume the efficiency of an process is given by $E = \int_0^T G(x(t))dt$, where the gain $G = \alpha x + \beta x^2$. Further we assume that the dynamics of the system is given by $x(t) = a_0 + a_1 \sin \omega t$, where the mean value a_0 is given and a_1 can be adjusted within some limits $0 \leq a_1 \leq a_0$. If $\beta = 0$ a stationary process with $a_1 = 0$ is optimal, i.e. $\partial E/\partial a_1 = 0$ for $a_1 = 0$. However, for $\beta > 0$, the oscillating state with $a_1 = a_0$ is most efficient. Many industrial processes have a such a nonlinear efficiency. Therefore a inhomogeneous chemical reaction with chaotic oscillations may be much more efficient than the corresponding stationary reaction. One reason for this gap between industrial needs and academic research is the lack of appropriate methods to distinguish between similar disordered materials and to characterize their properties. For example, it is well known from numerical simulations and theoretical investigations (FRENKEL, KONTOROVA [22]), that a one dimensional layer of atoms on top of a crystalline structure has a well defined aperiodic structure which depends on the size of the atoms compared to the periodicity of the underlying crystalline structure and the boundary conditions. As far as we know, nobody has tried to characterize those structures mathematically, in particular in higher dimensions, probably because it was nearly impossible to distinguish between them with spectroscopic methods.

This may change in the near future. In section 4 we present simple methods which make it possible to get a sharp distinction between similar disordered materials experimentally. In the next section we discuss methods to find structures emerging in high dimensional systems. If those methods are applied to nonequilibrium systems it might be possible to extract a quantitative relation between complexity and nonequilibrium with an appropriate definition of complexity.

3. SIMPLE MODELS FOR COMPLEX SYSTEMS

3.1 Slaving, Aims, Order Parameters, Evolving Systems

Synergetics and the theory on dynamical systems are concerned with the long-term behavior of the solutions of time dependent systems; in particular, the aim is to determine the level of self organization or chaos that the system can generate. Since the basic methodologies of synergetics and dynamical systems theory are well known and have been applied to different problems [23], here only the main results are recalled, in order to illustrate the differences between the approach in the present paper and the usual method.

A dynamical system in continuous time is a differential system,

$$d\tilde{x}(t)/dt = f(\tilde{x}(t), \tilde{p}(t) + \tilde{F}(t)), \tag{1}$$

where t is the time variable and $\tilde{x}(0) = \tilde{x}_0$. $\tilde{p}$ is a set of control parameters. Usually, some of those control parameters are experimentally accessible, i.e. can be changed in the course of an experiment. This is indicated by $\tilde{F}(t)$. $\tilde{p}$ and $\tilde{F}$ have the same dimension, but in practice most of the components of $\tilde{F}$ may be zero. $\tilde{x}(t)$ represents the actual state of the system. The corresponding state space can be finite or infinite dimensional. In a chemical system the components of $\tilde{x}$ might represent the concentrations of several species as well as temperature or the flow of heat at a certain location in the container. In a physical system, for example, a simple pendulum made from a metal bar which is fixed on one end and oscillates under the influence of a gravitational field, the components of $\tilde{x}$ might represent the motion of the center of mass of the bar and all vibrational states of the bar. For some dissipative systems it has been illustrated that all solutions of Eq.1 converge to a compact invariant set, the global attractor. For the pendulum we mentioned before, as well as for most other macroscopic mechanical setups, this means that all vibrational modes are slaved, i.e. decay rapidly to a limiting value which is uniquely determined by the motion of the order parameters, in this case the motion of the center of mass. For example, if we start out from a 2-dimensional system with second order nonlinearities but no driving forces $\tilde{F}(t)$:

$$\dot{\tilde{x}}_1(t) = \tilde{p}_1 \tilde{x}_1(t) - \tilde{x}_1(t)\tilde{x}_2(t) \tag{2}$$

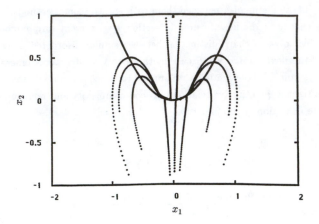

Figure 2: *The trajectories given by Eq.2 (doted lines) approach quickly the AIM (continuous line) given by Eq.5, where they advance slowly. The distance between the dots indicates the speed of the trajectory. ($\tilde{p}_1 = 0.01, \tilde{p}_2 = 1.0$)*

$$\dot{\tilde{x}}_2(t) \quad = \quad -\tilde{p}_2\tilde{x}_2(t) + \tilde{x}_1^2(t) \tag{3}$$

where $\tilde{p}_1$ and $\tilde{p}_2$ are two control parameters, and assume that $\tilde{x}_1(t)$ is slow compared to $\tilde{x}_2(t)$, i.e. $\tilde{p}_1 \ll \tilde{p}_2 \approx 1$ then one can solve Eq.3 by assuming that $\tilde{x}_1 \approx const.$:

$$\tilde{x}_2(t) = \frac{\tilde{x}_1^2(t)}{\tilde{p}_2} + \frac{1}{\tilde{p}_2}(\tilde{x}_2(0) - \tilde{x}_1^2(t))e^{-\tilde{p}_2 t}. \tag{4}$$

In Eq.4, the first term describes a slow motion given by $\tilde{x}_1(t)$ and the second term a transient which rapidly decays. After this rapid decay of the transient, $\tilde{x}_2$ is entirely determined by $\tilde{x}_1$:

$$\tilde{x}_2 = \frac{1}{\tilde{p}_2}\tilde{x}_1^2, \tag{5}$$

$\tilde{x}_2$ is "slaved" by $\tilde{x}_1$[1]. Since Eq.5 is invertible it is not obvious who slaves whom. However a certain asymmetry is given by the fact that a small sudden change of $\tilde{x}_1$ decays slowly or persists, whereas any change of $\tilde{x}_2$ decays fast, by definition. In a geometrical representation(Fig.2), the trajectory converges rapidly to a one dimensional manifold given by Eq.5, and then performs a slow motion given by Eq.5 and Eq.2:

$$\dot{x}_1 = p_1 x_1 - \frac{1}{p_2}x_1^3 \tag{6}$$

where we neglected the ~ in order to label x_1 as an order parameter and p_1 and p_2 as significant control parameters. This procedure is called adiabatic elimination of fast variables. Synergetics provides a systematic approach to estimate the location of those attractive, low dimensional limiting sets(Eq.5), the approximate inertial manifolds (AIM) and to determine the motion of the state vector on the AIM(Eq.6). In the above example we did not include any driving forces. In section 4 we study the same system including driving forces.

In most physical experiments the significant control parameters are not exactly constant but change at a slow rate compared to the order parameters. An example is the temperature or humidity in a laboratory which might change slightly during the experiment. Usually an experimentalist tries to fix the significant control parameters, for example through a temperature control or by putting the experiment in a special container. But one can never fix all of those control parameters. If one of those control parameters is changing at a noticeable rate, we call the system evolving. In an evolving system, the location of the AIM as well as the flow on the AIM may change as a function of time. The distinction between slaved variables, order parameters, and time dependent control parameters in an evolving system is based on time scale and stability arguments: slaved variables approach rapidly a limiting value which is determined by the order parameters. The dynamics of the order parameters emerges from the structure of the experimental system and may be complicated, but has usually a much longer time scale than the slaved variables. The significant control parameters change, if at all, on an even longer time scale which is typically longer than the period of the experiment and are usually set by the experimentalist. There may be experimental systems where this classification does not fit.

Given that the dynamics of a dissipative system always collapses rapidly onto one of those inertial manifolds, several questions arise: (i) Is the AIM smooth? Eq.6 contains a third order nonlinearity whereas the original equation (Eq.2) was second order. A successive elimination of more and more slaved variables in a high dimensional systems might produce higher and higher nonlinearities and converge to a non-smooth flow vector field. (ii) How attractive is the AIM? (iii) Does the systems stay on the AIM when an external force is applied in order to control the system? (iv) What are the properties of the equation of motion of the state vector on the AIM. In particular is the motion deterministic?

These questions are very important from a practical point of view. If the AIM is not smooth at a certain location, and the state vector reaches this location, it could mean rapid change of the concentration of chemical species and, for example, an explosion. (iii) addresses the question of whether we can use simplified chemical rate equations or the motion of the center of mass of a macroscopic mechanical system for a control. And (iv) means for mechanical systems: Can we expect that the dynamics of macroscopic quantities such as, for example, the center of mass of a pendulum to be deterministic and smooth? We discuss this problem of determinism in the context of classical mechanics in the next section. In section 4 we discuss the issue of controlling complex systems based on equations on the AIM.

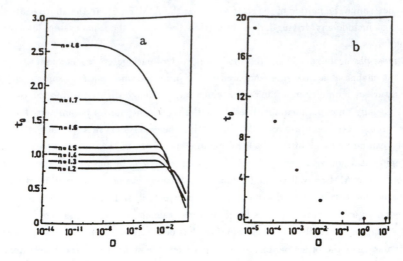

Figure 3: *This numerical study shows that the first passage time versus the noise level for several nondeterministic systems (a) has no singularity at zero noise level whereas a deterministic system (b) has a singularity.*

3.2 Nondeterminism and Free Will

Newton's principle of determinism states that the initial locations and velocities of a set of point masses uniquely determine all their future states[24]. The motion of macroscopic quantities of macroscopic mechanical systems is deterministic if both the AIM and the flow vector field on the AIM are smooth, which is generally not the case. The equations of motion of macroscopic mechanical systems are in general nonlinear, implicit, ordinary differential equations which can have unique solutions including deterministic chaos (type A) and nonunique solutions (type B). For type A chaos any future state is uniquely determined by the initial condition. However a slight change in the initial condition causes a large change of future states, whereas for nondeterministic systems (type B) future states may be undeterminable even if the initial state is exactly known and no noise is present. There are theorems on uniqueness for first order differential equations which satisfy a Lipschitz condition. There is no general proof that Euler's equations of motion for macroscopic mechanical systems can be made explicit in such a way that they satisfy a Lipschitz condition. The problem of uniqueness has to be studied case by case.

STELZL et al. and later DINKELACKER, SHERMER and others[25] studied a sphere which is in castor oil slowly sliding down a special slope which was made very accurately with a computer controlled milling machine. The equation of motion of this mechanical

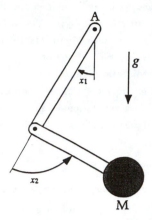

Figure 4: *The motion of a double pendulum in a gravitational field. M indicates the center of mass of the outer pendulum. When M comes close to A, the inertial moment may be very small. In this case large accelerations cause large forces, which stimulate a variety of complicated vibrations in a real setup.*

system reads:

$$\dot{x} = c|x|^{n-1} + F(t), \tag{7}$$

where $F(t)$ is uncorrelated white noise with standard deviation D. For $D = 0$ and $1 < n < 2$ the trajectories of this system intersect at $x = 0$ (type B dynamics), whereas for $n > 2$ the solutions are unique (type A dynamics). The system has a saddle point at $x = 0$ for all $n > 1$. However, in case of type B dynamics the mechanical system reaches the saddle point in a finite period of time, whereas for type A systems the ball gets closer and closer to the saddle point without reaching it exactly. It turns out that in the presence of noise statistical properties of type A and type B systems are very different. If started at a certain initial condition, the deterministic system passes the saddle point when the noise amplitude approximately equals the distance from the saddle point, i.e. the smaller the noise, the longer it takes to pass the saddle point. Fig.3 illustrates that the mean first passage time increases when the noise level is decreased for $x(0) = -0.78$ and $x_{final} = 0.0$ in type A systems. The situation is quite different in case of a type B system. In this case the trajectory reaches the saddle point after a finite period of time. Then the smallest kick is enough to pass the saddle. The mean first passage time is limited for small noise levels in this case. Therefore it should be possible to distinguish between deterministic and nondeterministic systems experimentally. This approach could be used, for example, to determine experimentally whether 3-dim Navier Stokes flows are deterministic or not.

There are strong indications that for most experimental systems the motion of the

AIM is not smooth, but discontinuous and nondeterministic. Even if the dynamics of the full system (Eq.1) is deterministic, and even if the neglect of some rapidly decaying variables is well justified through time scale arguments, there is no reason to assume that the property of determinism is preserved, if one or several variables in the equation of motion are neglected through adiabatic elimination.

To illustrate this conjecture, we present three examples. The equation of motion of a planar, <u>double pendulum</u> in a gravitational field can be found in any textbook on theoretical mechanics(Fig.4). If the mass of the upper pendulum is very small the equation of motion reads:

$$\ddot{x}_1 \ (\cos^2 x_2 - 1) =$$
$$(-(\dot{x}_1 - \dot{x}_2)^2 + \dot{x}_1^2 \sin x_2) \sin x_2 + g(\sin x_1 - \sin(x_1 - x_2) \cos x_2) \tag{8}$$
$$\ddot{x}_2 \ (\cos x_2 - 1) = (\dot{x}_1^2 - (\dot{x}_1 - \dot{x}_2)^2) \sin x_2 + g(\sin x_1 - sin(x_1 - x_2)) \tag{9}$$

where x_1 is the angle between the direction of the gravitational field and the axis of the upper pendulum, x_2 is the angle between the axes of the two pendula, and where g is the gravitational constant. The length of both pendula is assumed to be unity and the mass of the outer pendulum is assumed to be concentrated at its lower end and equals unity. These assumptions are not crucial for the point we want to make although it is crucial to note that these equations only describe the motion of the center of mass of the pendula, but not internal vibrations of each pendulum. Eq.8 describes the motion on an AIM where all internal vibrations are assumed to be slaved.

Whenever x_2 approaches π the term $(\cos x_2{}^2 - 1)$ at the left side of Eq.8 becomes zero and the trajectory passes through a singularity. From a mathematical point of view this singularity is removable, i.e. it is possible to find a continuous solution passing through the singularity. However, from a physical point of view this is not true, since the accelerations get very large at this state, which means strong forces are acting in the pendulum. In a real experiment these forces stimulate all kinds of internal vibrations and yield a sudden loss of energy of the macroscopic motion. This means that the AIM is not stable at this location and the macroscopic motion is not smooth. Even if the mass of the inner pendulum is not zero, as in all real experiments, forces tend to get very large at this state, since it is close to a singularity. Also, since a real double pendulum hits those states frequently it can be the main dissipation mechanism.

A second example is the motion of a <u>flag</u>. Let us assume that the front part of the flag is at rest whereas the rear end is bent around and moving with a velocity v_0(Fig.5). If we neglect all vibrational states of the flag we get a simple model which represents the conservation of energy:

$$m\dot{x}^2/2 = const \tag{10}$$

where x is the center of mass of the moving part of the flag, $m = (2/3)\rho(l - x)$ the mass

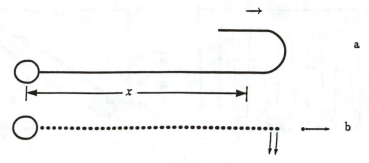

Figure 5: *The motion of a flag. Illustrated is a view from the top, where the circle indicates the post where the flag is fixed. Only the rear on the flag which is bent around is assumed to be moving (a). (b) illustrates the internal degrees of freedom of the flag and the rupture of the flag at the rear end. This process limits the life time of flags.*

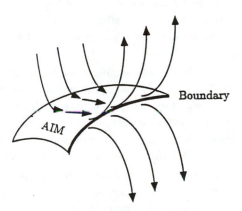

Figure 6: *The motion of trajectories at the boundary of an AIM.*

of the moving part, ρ the mass density and l the length of the flag. If we rewrite Eq.10:

$$\dot{x} = \sqrt{\frac{3 * const}{\rho(l-x)}} \tag{11}$$

we find that the solution of this differential equation ends at $x = l$. What happens in the real systems? All kinetic energy is stored at the very rear end of the flag and

16

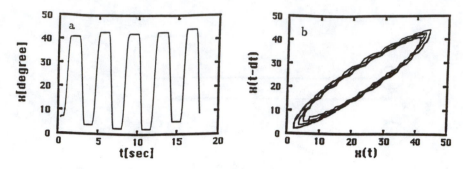

Figure 7: *The natural dynamics of a human arm location. (a) shows the angle of the arm versus time. The kink of the trajectory in the state space plot (b) indicates that the motion should be considered as nondeterministic (dt = 0.1sec).*

therefore the forces become so big that the very rear end of the flag breaks away. This means the trajectory on the AIM ended. Vibrational modes which had been slaved before get stimulated and can produce sound and damage the flag. The trajectory reached a boundary of the AIM(Fig.6).

The third example deals with the natural <u>motion of a human arm</u> or a human finger. Physiologists appreciate these systems because they provide quite reproducible results and seem to be low dimensional. Fig.7 shows an experimental time series of the natural motion of a human arm and a state space representation of the same time series. It has been shown that oscillators from the type of a van der Pol Oscillator $\ddot{x}+\eta(x^2-1)\dot{x}+p_1x+p_2x^2+p_3x^3+.... = F(t)$ are well suited to describe the global geometry of the limit cycle and the response of those motions to external perturbations, F[26]. However for $F = 0$ the dynamics of those VDP oscillators is always periodic, whereas the natural motions of the finger and arm are not periodic. Fig.7 shows that the arm seems to rest at the most extreme locations and after a rather undetermined period of time strats to move again. In addition, the motion of the arm can easily be stopped at those locations. This means that the model of a VDP oscillator does not adequately model the interference of free will with the motion of finger and arm. A careful analysis of the trajectories shows[27], that they have a kink at the most extreme locations(see Fig7b). A model which uses one VDP for the upper half plane of the state space and another VDP for the lower can perfectly model their global geometry as well as the kink. However the corresponding motion is no longer deterministic at $\dot{x} = 0$. After a finite period of time the smallest control force will make the finger move. This "nondeterministic model" provides both simple, low dimensional, accurate, unique equation of motion and nondeterministic states highly sensitive to any external perturbation due to the influence of free will. This finding might be useful for designing the motion of robots.

Nondeterminism might improve models for other biological systems, too. For example, a replacement of analytic sigmoidal functions[2] by a nonlinearity which produces a type B dynamics, might provide neural nets which reach optimal weights after a finite training period. Whether the dynamics of the human brain should be considered as of type A or B is unknown. Whether the motion of fluid flow is deterministic (type A) or not is unknown, but could be determined with accurate modeling techniques (sect. 3.4) or the method discussed at the beginning of this paragraph. The western view of the world is based on causality and determinism spoiled with noise. An approach where randomness and undeterminism is spoiled a little bit by causality might be a more adequate approach for real systems. CRUTCHFIELD's modeling theory with ϵ-machines[28] fits perfectly to these ideas, since it is not based on the assumption of an underlying deterministic motion.

3.3 Deterministic Chaos and Fractal Strange Attractors

For many complex systems it seems to be true that the high dimensional system is attracted to a low dimensional AIM and visits only a small, bounded region of this manifold: it approaches regular or strange attractors. Usually the adiabatic elimination procedure gives highly nonlinear flow vector fields, but if the experimental trajectory visits only a very small bounded region of the AIM without singularities, then it is often possible to get a good approximation by an appropriate expansion with cut off, for example a Taylor series expansion where all third order and higher order terms are neglected around the center of mass of the trajectory or a singular point. In this case the model of the complex system is low dimensional and has only nonlinearities of low order(LL-system). Nonlinear Dynamics has shown that LL systems can exhibit irregular motion which is sensitive to initial conditions: deterministic chaos. Therefore, deterministic chaos is important in complex systems research. Since the flow vector field of LL systems is smooth, their trajectories have often a simple geometry, for example they are fractal. Though this fact is generally only conjectured, FEIGENBAUM[29] proved it for the logistic map dynamics. Therefore, fractal strange attractors and deterministic chaos can be found in complex systems, but may not be generic for real systems, as we illustrated in the last section. But why are the Feigenbaum scenario and fractal strange attractors found on so many physical systems? The reasons are historical. The Feigenbaum scenario is typical for a sinusoidal driving force resonantly perturbing a damped, linear oscillator with a third order nonlinearity and very small higher order nonlinearities. Spectroscopy with small sinusoidal driving forces is the main topic in many physics labs, and most investigated are systems weakly damped and weakly nonlinear because then the response to sinusoidal perturbations, i.e. the signal to noise ratio, is as large as possible. For this equipment and those systems the Feigenbaum scenario is a generic route to chaos when the amplitude of the sinusoidal perturbation is increased.

3.4 Dynamical Reconstruction

The paradigm of dynamical reconstruction is the inverse of the paradigm of deterministic chaos. In this case a complicated time series is given and possible sources for such a time series are sought. In addition, we have to deal with the problem that in most experimental situations not all order parameters are observable. The paradigm on dynamical reconstruction states that *the most accurate model for prediction and control is obtained if the dynamics of the investigated system is highly chaotic and all preknowledge about the system is used, where it is unimportant whether the chaotic dynamics is imposed via external perturbations or is natural.*

From the point of view of an experimentalist the desired features of reconstructed models are:

- Simplicity: The model should encode an observed time series or spatial structure out of N data, with only a few parameters K, i.e. $K \ll N$, where the encoding itself should not be too complicated.

- Reproducibility: Various runs of the experiment must reproduce the parameters of the model.

- Physical Meaning of the Parameters: There should be a smooth relation, between the significant control parameters of the experiment and the parameters of the model in order to associate the parameters of the model with a physical meaning. In addition, a smooth relation between model parameters and experimentally significant control parameters is useful for predicting evolving systems.

- Prediction: The model should predict the future [30, 31] up to a certain accuracy and predict the dynamics for neighboring settings of the significant control parameters [32] and initial conditions.

- Accuracy and significance of the parameters: The neglect of a parameter of small magnitude or the neglect of a parameter which is smaller than its error bar should not decrease the quality of the model [33].

- Estimation of hidden order parameters: In many experimental systems not all order parameters are experimentally observable. The model should provide an estimate for those hidden order parameters.

Since noise affects the experimental system, it is never possible to reproduce the parameters of the equation of motion exactly. A measurement is considered to be reproducible if statistical tests indicate that the deviations of the parameters between individual measurements are due to the influence of noise. We will discuss such a test.

The modeling may be considered as a special type of filter for the data. If the dynamics of an experimental system is periodic, it turns out that a fast Fourier transformation very frequently satisfies all those conditions with just a few nonzero Fourier components. These Fourier amplitudes together with the corresponding Fourier frequencies are the parameters of a model. But as soon as transient dynamics, resulting from an initial condition which does not belong to the limit cycle, is included in the time series, the dynamics is no longer periodic, i.e. the Fourier spectrum gets complicated and the number of nonzero parameters may equal the number of points in the time series [35]. In particular, if the limiting dynamics is chaotic, nearly all those parameters get large. Since in this case the condition $K \ll N$ is violated, a Fourier transformation is not an appropriate modeling process for the chaotic time series.

Assume the observed time series originates from a system whose dynamics can be described by an N_d-dimensional system of differential equations of first order $\dot{\vec{y}} = \vec{f}(\vec{y})$ ($\vec{y} \in S = N_d$-dimensional state space [36, 30]) or from a map $\vec{y}_{n+1} = \vec{f}(\vec{y}_n)$ ($n = 1, 2, \ldots$) with a flow vector field absent of singularities. Then we can model the dynamics by an analytical estimation of its flow vector field. In certain cases this analytical estimate of the flow vector field may have just a few fit parameters. In this case the analytical estimate of the flow vector field is a simple model of the aperiodic time series. In the following we will focus on this situation.

The problem of dealing with hidden variables in dynamical systems was first addressed by PACKARD et al.[36] and RUELLE in 1980. They noted that a state space representation of the dynamics could be reconstructed from a single time series through the use of delay coordinates. This delay-coordinate reconstruction would then be topologically equivalent to the dynamics of the true system. WHITNEY had shown much earlier that any compact manifold with dimension m can be embedded in R^{2m+1}[37]. TAKENS extended this in proving that an embedding can be obtained from any system from only a time series by using $2m + 1$ delay coordinates[38]. While this combination of ideas thus far has been extremely useful in studying nonlinear systems, several difficulties arise in their application.

The most obvious difficulty in using delay coordinates is the issue of interpreting the results physically. If one is concerned only with forecasting, the method used for the modeling is irrelevant, as long as it works. Successful modeling techniques based upon delay coordinates and/or partitioning the state space to generate local fits have been developed[30, 39]. However, relating these models to physical principles or existing theories is often difficult if not impossible. For example if the two variables of a three dimensional ODE are hidden, and the time series of the third variable is represented in state space using delay coordinates, the simplest model for the flow is typically an integro-differential equation, i.e. substantially more complicated than the ODE. Another important consideration is that Takens's theorem does not apply to systems with noise, i.e. experimental data. And Whitney's theorem as well as Takens' modification apply only to smooth and stationary manifolds, whereas smooth and stationary AIMs may not be generic in experimental situations.

The dynamical reconstruction[40] which we present in the next few paragraphs addresses the problems associated with an optimal state space reconstruction as well as desirable features for the modeling process. To illustrate the basic idea we use a simple example. Given is a system of two coupled maps:

$$x_1(t+1) = f(x_1(t), x_2(t), p_1 + F_{noise}^1(t) + F_1(t)) \tag{12}$$

$$x_2(t+1) = f(x_1(t), x_2(t), p_2 + F_{noise}^2(t) + F_2(t)) \tag{13}$$

where $x_1(t), t = 1, 2, \ldots$ is assumed to be measurable and given, whereas $x_2(t)$ is unobservable, i.e. a hidden variable. p_1 and p_2 are two sets of parameters, for example, the coefficients of a Fourier expansion of f. $F_{noise}^1(t)$ and $F_{noise}^2(t)$ represent small uncorrelated noise and F_1 and F_2 are control forces. F_1 and F_2 are assumed to be known and may be zero. If p_1 and p_2 are known, we use a maximum likelihood method to estimate the value of the hidden variable[41], i.e. $x_2^{guess}(t)$ is estimated by the x_2 value of the most probable trajectory of length i_{max}:

$$x_2^{guess}(t) \in \{\hat{x}_2^o(t) | \prod_{i=0}^{i_{max}} P(t, t_i, t_{i+1}, \hat{x}_2^o(t+t_i), p_1, p_2)$$

$$\geq \prod_{i=0}^{i_{max}} P(t, t_i, t_{i+1}, \hat{x}_2(t+t_i), p_1, p_2)\} \tag{14}$$

where $t_0 = 0$, $t_{i+1} > t_i$, $\hat{x}_2(t+1) = f(x_1(t), \hat{x}_2(t), p_2 + F_2(t))$, and $\hat{x}_2^o(t+1) = f(x_1(t), \hat{x}_2^o(t), p_2 + F_2(t))$. $P(t, t_i, t_{i+1}, x_2, p_1, p_2)$ is the probability of the transition from the state $(x_1(t+t_i), x_2)$ to a state with a given $x_1(t+t_{i+1})$ within $\Delta t = t_{i+1} - t_i$ steps in time. In principle one may choose $t_{i+1} = t_i + 1$ with a large i_{max}, but if x_2 is estimated numerically this would need a lot of computer power. Systematic numerical studies indicate that the estimate for x_2 does not substantially degrade if $i_{max} > 1$ is small, for example equal to two, and if Δt equals a quarter of the typical time scale of the system. If x_2^{guess} cannot be determined uniquely it is sometimes useful to increase i_{max}.

If P is Gaussian, the optimum value is given by the minimum value of Q, the mean square difference between the predicted and the observed time series $Q(t, \hat{x}_2(t), p_1, p_2) = \sum_{i=1}^{i_{max}} (x_1(t+t_i) - f_1(x_1(t+t_i-1), \hat{x}_2(t+t_i-1), p_1 + F_1(t+t_i-1)))^2$:

$$x_2^{guess}(t) \in \{\hat{x}_2^o(t) | Q(t, \hat{x}_2^o(t), p_1, p_2) \leq Q_j(t, x_2(t), p_1, p_2)\} \tag{15}$$

If only rough estimates of p_1 and p_2 are known, they can be improved by a maximum likelihood estimation, too:

$$(p_1, p_2)^{opt} \in$$
$$\{(p_1^o, p_2^o) | \prod_{i=0, j=1}^{i_{max}, j_{max}} P(t_j, t_i, t_{i+1}, x_2^{guess}(t_j), p_1^o, p_2^o)$$
$$\geq \prod_{i=0, j=1}^{i_{max}, j_{max}} P(t_j, t_i, t_{i+1}, x_2^{guess}(t_j), p_1, p_2)\} \tag{16}$$

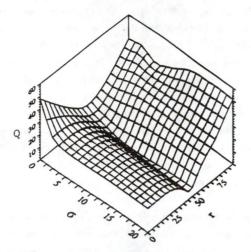

Figure 8: *Q versus the parameters of the model, σ and r where the time series $x_1(t)$, $x_3(t)$ and the structure of the equation: $\dot{x}_1 = \sigma(x_2 - x_1), \dot{x}_2 = -rx_1 - x_2 - x_1 x_3, \dot{x}_3 = -8/3x_3 + x_1 x_2$ are given. $x_2(t)$, σ, and r are unknown. Q has a minimum at the exact parameter settings. For details see [40].*

and if P is Gaussian through a minimization of Q:

$$(p_1, p_2)^{opt} \in \{(p_1^o, p_2^o)| \sum_{j=1}^{j_{max}} Q(t_j, x_2^{guess}(t_j), p_1^o, p_2^o)$$

$$= \min \sum_{j=1}^{j_{max}} Q(t_j, x_2^{guess}(t_j), p_1^o, p_2^o)\} \tag{17}$$

where t_j is a set of times, for example $t_j = 2j$. In section 4.2 we illustrate how to estimate Q algebraically for the case where all order parameters are observable. It seems that this approach can be generalized to the modeling process with hidden variables.

There are many methods to search for the extrema of the likelihood (Eq.16) or Q (Eq.16) numerically(Fig.8). A simple approach is to search the minimum with a gradient search and an appropriate choice of initial conditions. As an initial condition one picks the rough estimate and a certain number of neighboring positions in order to check for local minima. In particular if the system is evolving, i.e. $p_1 = p_1(t)$ and $p_2 = p_2(t)$, then the estimates from the previous modeling process are well suited as initial conditions. For evolving systems only data from the most recent history should be used for the modeling. A detailed discussion of this issue was given by OHLE et al.[33]. If no rough estimates for p_1 and p_2 are available, one has to define a region of interest and start the search from a large number of initial conditions within this region.

There is never a unique model for the whole state space, since it is impossible to collect data from all regions of the state space. However, it may be possible to cover a certain region of the state space with data. If the measured variables and the hidden variables cover this region of interest then the model parameters as well as their accuracy can be determined uniquely[33]. The parameters of the model are most accurate if the dynamics covers this region homogeneously[34]. In order to obtain such a homogeneous coverage, intrinsic noise in the system as well as external driving forces, which force the system to cover the region of interest homogeneously may be useful[40] . Therefore it is advantageous for the modeling process to impose a random motion or a highly chaotic motion onto the experimental system. However, there is a big problem associated with such a control: slaved variables may be stimulated and make the reconstruction of a simple model impossible. To do a control without stimulating slaved variables is the main issue of dynamical control and the principle of the dynamical key, which we will discuss later.

It turns out that the proposed modeling process optimizes most of the requirements which we mentioned above. It is not confined to smooth dynamics on the AIMs or smooth AIMs since f in Eq.12 may be singular or even multivalued. The obtained models tend to be simple, since foreknowledge about the system can be used. Since the model parameters can be determined uniquely, including their error bars, there are usually no problems with reproducibility or interpolations between neighboring parameter settings. In the next few paragraphs we discuss how to use such models for a control.

4. CONTROLLING COMPLEX SYSTEMS

4.1 Nonlinear Resonances and the Principle of the Dynamical Key

The principle of the dynamical key states: *Nonlinear systems react most sensitively to a special class of aperiodic driving forces. Typically, the characteristic time scales of the optimal driving force match at all times the characteristic time scales of the system. In some cases the optimal driving force as well as the resulting dynamics are similar to the transients of the unperturbed system.*

The following theorem illustrates this paradigm in a physical context[42]. Given is the velocity of the oscillation $\dot{x}(t)$ and the driving force $F(t)$ in terms of amplitude $A(t) \geq 0$, $B(t) \geq 0$, time scale $T_x(t) = 2\pi/|\omega_x(t)|$, $T_F(t) = 2\pi/|\omega_F(t)|$ and phases ϕ_x, ϕ_F:

$$\dot{x}(t) = A(t)\sin\left(\omega_x(t) * t + \phi_x\right) \tag{18}$$

$$F(t) = B(t)\sin\left(\omega_F(t) * t + \phi_F\right) \tag{19}$$

where $x(t)$ is a state vector and $\omega_x(t)$ and $\omega_F(t)$ are real or imaginary scalars. There is a a 100% energy transfer $P(t)$, i.e. 0%of the transferred energy are reflected and

$$P(t) = \dot{x}(t)F(t) \geq 0 \tag{20}$$

for all times t, if the two time scales and phases are equal. In this theorem we used the efficiency of the energy transfer as a measure of the size of the response. This definition is useful for physical and chemical systems and might help to understand phenomena such as catalysis[50]. But the paradigm applies for other measures, too. We will show some examples later. There is a certain arbitrariness in the definition of ω_x and ω_F. Since the theorem applies for any definition of ω_x and ω_F, it means that all time scales have to match. However, it provides useful physical interpretations only if the definitions are appropriately chosen for a given dynamical system. Another critical question concerns the type of similarity between the driving force and the transient dynamics needed to get an optimal response. No general answer is known yet, but we will give some examples. In the next few paragraphs we compare the impact of sinusoidal driving forces and optimal driving forces on low dimensional nonlinear oscillators and complex systems and study some examples.

Sinusoidally driven nonlinear oscillators have complex, in many cases chaotic, dynamics. The response to externally applied forces is relatively small and does not match any well defined resonance condition. E.g., when a nonlinear mechanical pendulum is perturbed by a sinusoidal force the response is comparatively small in amplitude[42], is in many cases chaotic[43] and does not satisfy any well defined resonance condition, even when the frequency of the driving force coincides with a peak (resonance) in the power spectrum of the dynamics of the unperturbed system[44]. In order to obtain a large response, the frequency of the driving force must be shifted in such a way that it coincides at all amplitudes with the characteristic frequency of the oscillator[42]. But even when the frequency of the driving force coincides at all amplitudes with the basic resonance of the unperturbed system, the resonance condition is not exactly satisfied, since all the other peaks[44] of the power spectrum of the unperturbed system have to be taken into account as well. Recently a method has been presented to calculate driving forces which satisfy the resonance condition exactly[42]. We will show that they are in agreement with the principle of the dynamical key. In general, resonant driving forces are those which reach a certain target, break a special bonding of a molecule, or satisfy a given constraint with the smallest driving force possible.

Perturbations of small amplitude are most important for the stimulation of complex systems. Its has been shown[1] that due to the slaving principle a large variety of complex systems can be modeled by low dimensional systems of differential equations. If one uses this system of differential equations to calculate driving forces which force the experimental system to perform a special dynamics (goal dynamics) and applies these forces to the real system, then generally some slaved variables which are not included in the model will be excited, too. If we take for example the system given by Eq.2 and investigate the influence of an additive driving force we would expect that the response is given by:

$$\dot{x}_1 = p_1 x_1 - p_2 x_1^3 + F(t). \tag{21}$$

In order to check this expectation we go back to the full set of equations and assume, for the sake of simplicity, that the driving force $F(t)$ acts on both variables in the same way:

$$\dot{\tilde{x}}_1(t) = \tilde{p}_1\tilde{x}_1(t) - \tilde{x}_1(t)\tilde{x}_2(t) + \tilde{F}(t) \tag{22}$$

$$\dot{\tilde{x}}_2(t) = -\tilde{p}_2\tilde{x}_2(t) + \tilde{x}_1^2(t) + \tilde{F}(t) \tag{23}$$

where $\tilde{p}_1$ and $\tilde{p}_2$ are two control parameters. If we assume that $\tilde{x}_1(t)$ is slow compared to $\tilde{x}_2(t)$, i.e. $\tilde{p}_1 \ll \tilde{p}_2 \approx 1$, then one can solve Eq.3, by assuming that $\tilde{x}_1 \approx const.$ and $F(t) \approx const$ in Eq.23:

$$\tilde{x}_2(t) = \frac{1}{\tilde{p}_2}\tilde{x}_1^2(t) + \tilde{F}(t) + (\tilde{x}_2(0) - \frac{1}{\tilde{p}_2}(\tilde{x}_1^2(t) + \tilde{F}(t)))e^{-\tilde{p}_2 t} \tag{24}$$

In Eq.24 the first two terms describe a slow motion given by $x_1(t)$ under the influence of a driving force and the third term a transient which rapidly decays. After this rapid decay of the transient, $\tilde{x}_2$ is entirely determined by $\tilde{x}_1$ and $\tilde{F}$:

$$\tilde{x}_2 = \frac{1}{\tilde{p}_2}\tilde{x}_1^2 + \tilde{F}, \tag{25}$$

$\tilde{x}_2$ is "slaved" by $\tilde{x}_1$ and $\tilde{F}$. In a geometrical representation, the trajectory converges rapidly to a two dimensional manifold given by Eq.25 and then performs a slow motion given by Eq.25 and Eq. 22:

$$\dot{x}_1 = (p_1 + F(t))x_1 - \frac{1}{p_2}x_1^3 + F(t). \tag{26}$$

where we neglected the $\tilde{}$ in order to label x_1 as a order parameter. It turns out that Eq.26 is in agreement with Eq.21 if we assume that $F(t) \ll p_1 \ll p_2$. The result of this little illustration is that F has to be very small and slow, whereas there is no constraint on the size of the response. Very similar results can be obtained for a macroscopic physical pendulum. If an external force is applied to the pendulum, the response of the pendulum is given by the equation of motion of the center of mass which contains the driving force as an additive term. However, if the driving force is large and has a short time scale, as for example a kick by a hammer(Fig.9), this equation does not give a correct description of the response, since all kinds of vibrational modes will be stimulated. Only if the typical time scale of the driving force is much larger than the typical time scale of all slaved variables can the response of the pendulum be predicted with the low dimensional model.

Usually, if the driving force is small, the response will be small, too, unless the driving force is resonant. But resonant perturbations have another advantage besides their large responses and small amplitudes: one can predict the response of a high dimensional system with a low dimensional system of differential equations, since excitations aimed away from the inertial manifold generally remain small. We will show that these resonant perturbations are aperiodic. In the next paragraph we show that in contrast to optimal aperiodic driving forces periodic driving forces do not produce a large response, even if they are at resonance.

A general feature of nonlinear oscillators is the amplitude frequency coupling[45]. The stimulation of a nonlinear oscillator is *at resonance* if the perturbation is periodic and

Figure 9: *If a physical pendulum is stimulated with a hammer instead of a resonant driving force slaved vibrations get excited.*

if the frequency of the perturbation coincides with the basic frequency of the oscillator. This means for a driving force *at resonance* the initial conditions satisfy the principle of the dynamical key. However, as soon as the first energy transfer happens, it is violated because of the amplitude frequency coupling. In order to illustrate that the response *at resonance* is small compared to the response to special nonsinusoidal perturbations which will be presented later in this section, we investigate Hamiltonians of the form

$$H = V_1\left(x_1\right) + \frac{1}{2}m_1\dot{x}_1^2 + V_2\left(x_2\right) + \frac{1}{2}m_2\dot{x}_2^2 + \frac{1}{2}k\left(x_1 - x_2\right) \tag{27}$$

where x_1 and x_2 are amplitudes, m_1 and m_2 are inertial constants, V_1 is an anharmonic and V_2 a harmonic potential and k is a small coupling constant which quantifies the magnitude of the coupling between the harmonic driving system and the nonlinear oscillator. In order to keep the feedback from the nonlinear oscillator to the dynamics of the driving system as small as possible, we assume that the energy of the driving system is considerably larger than the energy of the nonlinear oscillator. In order to quantify the magnitude of the response the maximal energy transfer to the nonlinear oscillator is used. Fig.10 illustrates a typical resonance curve.

The area of the resonance peak estimated by $A = \Delta\omega\Delta E_{max}$, where $\Delta\omega$ is the width of the resonance curve and ΔE_{max} is the height of the resonance curve, is independent of the magnitude of the amplitude frequency coupling if $\Delta\omega$ and ΔE_{max} are estimated by a first order secular perturbation theory. Using the same approximation, one can show[47] that there is a simple relation between the quality factor of the resonance estimated by

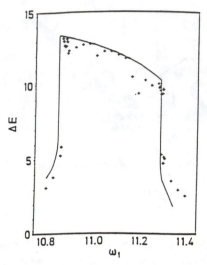

Figure 10: *Typical resonance curve of a nonlinear oscillator, where $V_1 = x_1^6, m_1 = m_2 = 1, k = 0.1$, and the energy of the harmonic oscillator $E_2 \approx 100$ is considerably larger than the energy of the nonlinear oscillator $E_1 \approx 10$. Plotted is the maximal energy transfer versus the frequency of the perturbation where the initial energy of the nonlinear oscillator was kept fixed. The continuous line represents the results of secular perturbation theory and (+) represents numerical results[46]. Generally those resonance curves have a flat top region and sharp edges.*

$Q = \frac{\Delta E_{max}}{\Delta \omega}$ and the magnitude of the amplitude frequency coupling

$$Q = \frac{1}{\frac{\partial \omega_1}{\partial E_1}} \tag{28}$$

where ω_1 is the basic frequency of the nonlinear oscillator for $k = 0$ and which depends on the energy E_1 of this oscillator. Since the area of the resonance peak is independent of the amplitude frequency coupling, we conclude from Eq. (28) that the resonance curve becomes broad and small, if the driven oscillator is considerably nonlinear, i.e. has a large amplitude frequency coupling.

In order to calculate resonant driving forces $F(t)$ for nonlinear systems we start out with a damped motion in a nonlinear potential:

$$m\ddot{x} + \eta_1 \dot{x} + \frac{\partial V(x)}{\partial x} = F(t) \tag{29}$$

where m is the mass, η_1 a friction coefficient and where the first term represents inertial force, the second term a velocity dependent friction force, the third term a nonlinear potential force and $F(t)$ a driving force.

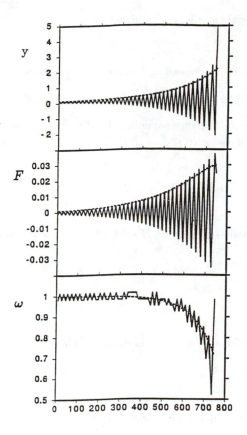

Figure 11: *The goal dynamics y, the optimal driving force, and its frequency versus time (—-) for a fixed η_2 for a damped physical pendulum. Depicted is every third extremum. They are connected with lines. We calculate first the goal dynamics, then use the resulting time series for the calculation of F and finally integrate Eq.29. y(t) provides also an illustration of x(t), since after a short transient at t = 0, x is equal to y within the numerical accuracy. The numerical estimate of ω results from the period between succeeding extrema of y(t). (- - -) shows the analytic estimates for the amplitudes of the three quantities[17].*

A strong resonance is defined through the following set of conditions: $x(t = 0) = x_0$, $x(t = T) = x_T$, T, $H(t = 0) = H_0$, $H(t = T) = H_T$ where $H = \dot{x}^2/2 + V(x)$ are given, i.e.

$$\Delta H = \int_0^T F\dot{x} - \eta_1 \dot{x}^2 dt = fixed \qquad (30)$$

and where the mean square average of the driving force

$$\bar{F^2} = \int_0^T F(t)^2 dt \qquad (31)$$

is as small as possible. The corresponding Lagrange equation reads:

$$L = F(t)^2 - \lambda_1(t)(m\ddot{x} + \eta_1\dot{x} + \frac{\partial V(x)}{\partial x} - F(t)) - \lambda_2(F\dot{x} - \eta_1\dot{x}^2) \qquad (32)$$

where $\lambda_1(t)$ and λ_2 are the Lagrange parameters for the constraints given by Eq.29 and Eq.30. The equations of motion $\partial L/\partial x - d/dt\partial L/\partial\dot{x} + d^2/dt^2\partial L/\partial\ddot{x} = 0$ and $\partial L/\partial F = 0$ yield the equations

$$-\lambda_1\frac{\partial^2 V(x)}{\partial x^2} + \eta_1\dot{\lambda}_1 + \lambda_2\dot{F} - 2\lambda_2\eta_1\ddot{x} - \ddot{\lambda}_1 = 0 \qquad (33)$$

$$2F + \lambda_1 - m\lambda_2\dot{x} = 0 \qquad (34)$$

in addition to the constraints given by Eq.29 and Eq.30. A solution of these four equations is given by

$$F(t) = 2\eta_1\dot{y} \qquad (35)$$

$$\lambda_1(t) = -4\eta_1\dot{y} \qquad (36)$$

$$\lambda_2 = 0 \qquad (37)$$

where the dynamics of the auxiliary variable $y(t)$ if given by

$$m\ddot{y} - \eta_1\dot{y} + \frac{\partial V(y)}{\partial y} = 0 \qquad (38)$$

Generally $y(t)$ is not periodic if $\frac{\partial V(y)}{\partial y}$ is nonlinear. Therefore the optimal driving force $F(t)$ is aperiodic. In addition, the dynamics of $x(t)$ is given by

$$x(t) = y(t) \qquad (39)$$

For many systems the solution given in Eq.39 is stable and has a certain basin of attraction. In this case Eq.39 provides an estimate of the response $x(t)$ even if $x(t = 0) \neq y(t = 0)$, but is in the basin of attraction. Since Eq.38 is autonomous, i.e. has no explicit time dependence, in contrast to Eq.29, it is sometimes possible to solve Eq.38 in a closed form. In this case the response due to the aperiodic, optimal driving force is given in a closed form[17], i.e. is simple in terms of standard algebraic concepts(Fig.11).

Eq. 38 results from unperturbed experimental dynamics (Eq. 29) through a reflection of time. This fact provides a physical interpretation: The nonlinear oscillator reacts most sensitive if perturbed by a driving force which mimics the time reflected transient dynamics of the unperturbed system(Fig.12). In this case the driving force matches at all times the actual rhythm of the nonlinear oscillation and the absorption of energy is 100%, i.e. the energy transfer is always positive(principle of the dynamical key):

$$P(t) = F(t)\dot{x}(t) = 2\eta_1\dot{x}^2(t) \geq 0 \qquad (40)$$

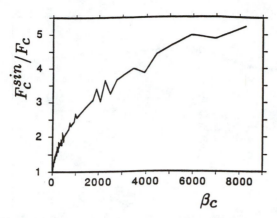

Figure 12: *The ratio of the critical amplitudes of the driving forces F_c/F_c^{sin} versus the Stewart-McCumber number β_c for $\rho = 0.1$, where x models the dynamics of the phase of a Josephson junction: $\ddot{x} + 1/\beta_c\dot{x} + \rho + \sin x = F(t)$. F_c and F_c^{sin} are the minimum amplitudes of aperiodic optimal driving forces (strong resonance) and sinusoidal driving forces respectively in order to switch the junction from a libration to a rotation[17]. For very small friction F_c/F_c^{sin} can be as large as 10^8[18, 48].*

This is an example of the similarity relation mentioned in the principle of the dynamical key: the response of the experimental systems equals the time reflected transient dynamics of the unperturbed system (Eq. 39)and the optimal driving force is proportional to the time derivative of the time reflected transient dynamics of the unperturbed system (Eq. 35). And, of course, the typical time scales of the driving force F and the velocity $\dot{x}$ match, since, for the solution given by Eq.39, they are directly proportional to each other: $F(t) = 2\eta_1\dot{y} = 2\eta_1\dot{x}$

If the experimental system is <u>conservative</u> the driving force given by Eq.35 is equal to zero. However in this case there is another solution of Eq.33:

$$F(t) = 2\frac{\dot{y}}{t+c} \qquad (41)$$

where the dynamics of the auxiliary variable $y(t)$ is given by

$$m\ddot{y} - 2\frac{\dot{y}}{t+c} + \frac{\partial V(y)}{\partial y} = 0 \qquad (42)$$

c is a constant. For $c < 0$, energy is extracted from the oscillator and for $c > 0$, energy is transferred to the oscillator. In addition, the dynamics of $x(t)$ is given by

$$x(t) = y(t) \qquad (43)$$

30

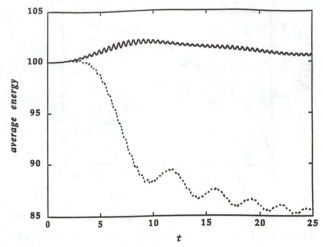

Figure 13: *Cooling of a conservative system* $(V(x) = 2.5x^4)$ *without feedback[49]. The initial energy content is* $H_0 = 100$, *the initial phase is randomly chosen. Plotted is the average energy versus time, for optimal cooling force with* $c = 50$, $\dot{y}(0) = 0$, *and* $y(0) = (H_0/2.5)^{1/4}$ *(dashed line) and for a sinusoidal driving force* $F(t) = 0.8\cos\omega_0 t$ *where* ω_0 *is the resonance frequency at* $t = 0$.

For many systems the solution given in Eq.39 is metastable. If $x(t = 0) \neq y(t = 0)$, but $x(t = 0) \approx y(t = 0)$ the difference between $x(t)$ and $y(t)$ tends to grow, which makes it impossible to do a successful control over a long period of time. However if $x(t = 0) \approx y(t = 0)$ it is in many cases possible to control the system successfully for short periods, and this may be long enough for successfully breaking the bond of a molecule (section 4.2) or to cool a conservative oscillator. If we take for example the conservative oscillator:

$$\ddot{x} + x^3 = F(t) \tag{44}$$

with a given initial energy content H_0, it is impossible to cool down the oscillator with a sinusoidal driving force $F(t) = A\sin\omega t$. For small A the energy content of the oscillator is confined to a small interval which is determined by the neighboring KAM curves. If an optimal driving force (Eq.41) is used, it is possible to extract a substantial amount of energy even if the initial phase of the oscillator is unknown(Fig.13). Due to the symmetry of periodic driving forces the sign of the energy transfer is determined by the initial phase which makes it impossible to extract energy intentionally if the initial phase is unknown. In contrast, optimal driving forces are aperiodic and the sign of the energy transfer is determined by the initial phase and frequency shift. Even if the initial phase is unknown, there is a tendency to extract energy or to store energy depending on the frequency shift. The larger the amplitude frequency coupling of the conservative oscillator the larger this effect will be[49].

A <u>medium resonance</u> is defined through the following set of conditions: $x(t = 0) = x_0$, $x(t = T) = x_T$, T, $H(t = 0) = H_0$, are given and

$$\Delta H = \int_0^T F\dot{x}dt = fixed \tag{45}$$

and where the mean square average of the driving force

$$\bar{F}^2 = \int_0^T F(t)^2 dt \tag{46}$$

is as small as possible. The corresponding Lagrange equation reads:

$$L = F(t)^2 - \lambda_1(t)(m\ddot{x} + \eta_1\dot{x} + \frac{\partial V(x)}{\partial x} - F(t)) - \lambda_2(F\dot{x}) \tag{47}$$

where $\lambda_1(t)$ and λ_2 are the Lagrange parameters for the constraints given by Eq.29 and Eq.45. The equations of motion $\partial L/\partial x - d/dt\partial L/\partial\dot{x} + d^2/dt^2\partial L/\partial\ddot{x} = 0$ and $\partial L/\partial F = 0$ yield the equations

$$-\lambda_1\frac{\partial^2 V(x)}{\partial x^2} + \eta_1\dot{\lambda}_1 + \lambda_2\dot{F} - \ddot{\lambda}_1 = 0 \tag{48}$$

$$2F + \lambda_1 - \lambda_2\dot{x} = 0 \tag{49}$$

in addition to the constraints given by Eq.29 and Eq.45. A solution of these four equations is given by

$$F(t) = (\eta_1 + \eta_2)\dot{y} \tag{50}$$

$$\lambda_1(t) = -\frac{(\eta_1 + \eta_2)^2}{\eta_1}\dot{y} \tag{51}$$

$$\lambda_2 = -\frac{(\eta_1 + \eta_2)^2}{\eta_1} + 2(\eta_1 + \eta_2) \tag{52}$$

where the dynamics of the auxiliary variable $y(t)$ if given by

$$m\ddot{y} - \eta_2\dot{y} + \frac{\partial V(y)}{\partial y} = 0 \tag{53}$$

In addition, the dynamics of $x(t)$ is given by

$$x(t) = y(t) \tag{54}$$

For many systems the solution given in Eq.54 is stable and has a certain basin of attraction. In this case Eq.54 provides an estimate of the response $x(t)$ even if $x(t = 0) \neq y(t = 0)$, but is in the basin of attraction. Since Eq.53 is autonomous, i.e. has no explicit time dependence, in contrast to Eq.29, it is sometimes possible to solve Eq.53 in a closed form. In this case the response due to the aperiodic, optimal driving force is given in a closed form[17], i.e. is simple in terms of standard algebraic concepts.

Again, the absorption of energy is 100%, i.e. the energy transfer always positive (principle of the dynamical key):

$$P(t) = F(t)\dot{x}(t) = 2\eta_1\dot{x}^2(t) \geq 0 \tag{55}$$

This is an example of a more complicated similarity relation mentioned in the principle of the dynamical key: the response of the experimental systems equals the time reflected transient dynamics of a system which has the same equation of motion as the unperturbed system (Eq. 54) but a with a different friction constant. The optimal driving force for medium resonance is proportional to the time derivative of the time reflected transient dynamics of the the the same system (Eq. 50). And, of course, the typical time scales of driving force F and the velocity $\dot{x}$ match since, for the solution given by Eq.54, they are directly proportional to each other: $F(t) = 2\eta_1 \dot{y} = 2\eta_1 \dot{x}$

A weak resonance is defined through the following set of conditions: $x(t = 0) = x_0$, $x(t = T) = x_T$, T, $H(t = 0) = H_0$, are given and where the reflected energy

$$H_r = \int_0^T F(t)\dot{x}\Theta(-F(t)\dot{x})dt \tag{56}$$

is as small as possible. A solution of these equations is given by

$$F(t) = (\eta_1 + \eta_2(t))\dot{y} \tag{57}$$

where the dynamics of the auxiliary variable $y(t)$ is given by

$$m\ddot{y} - \eta_{2(t)}\dot{y} + \frac{\partial V(y)}{\partial y} = 0 \tag{58}$$

where $\eta_2(t) \geq 0$. In addition, the dynamics of $x(t)$ is given by

$$x(t) = y(t) \tag{59}$$

For many systems the solution given in Eq.59 is stable and has a certain basin of attraction. In this case Eq.59 provides an estimate of the response $x(t)$ even if $x(t = 0) \neq y(t = 0)$, but is in the basin of attraction. Again, the absorption of energy is 100%, i.e. the energy transfer always positive.

This is an example of another more complicated similarity relation mentioned in the principle of the dynamical key: the response of the experimental systems equals the time reflected transient dynamics of a system which is equal to the unperturbed system (Eq. 59) except that the friction constant is different and time dependent. And the optimal driving force is proportional to the time derivative of the time reflected transient dynamics of the same system (Eq. 57). As before, the typical time scales of driving force F and the velocity $\dot{x}$ match since, for the solution given by Eq.54 they are directly proportional to each other: $F(t) = (\eta_1 + \eta_2(t))\dot{y} = (\eta_1 + \eta_2(t))\dot{x}$. A special case is the situation where $\eta_2(t) = F_{max}/(\dot{x}^2 + 2V(x))^{1/2}$ and where F_{max} is a given constant. D. BENSEN has shown[48], that in this case the maximum amplitude of the driving force is approximately constant and given by F_{max}[48].

4.2 Optimal Waves With the Dynamical Key

A generalization of these definitions of resonances to quantum systems and other spatially extended systems is straightforward. It has been applied to create solitons efficiently[18, 51]. In this paper we discuss two examples: (i) optimal dissociation of a diatomic molecule, which was studied by S. KREMPL et al.[52] and (ii) optimal diffraction patterns from aperiodic structures, which was investigated by R. WITTMAN, M. MONDELLO and others[53].

Optimal control of quantum systems has become a field of major interest in recent time[54]. Several theoretical works[54]·[55]·[56] have shown that, by using specially designed laser pulses, molecules could very successfully be excited in a way that would not be possible with single frequency lasers[57]. In the past few years techniques to generate ultrashort laser pulses[58]·[59] have been developed that allow precise amplitude and frequency control. In recent work, optimal fields have been derived from properties of two level systems[55], classical calculations[56] or by sophisticated numerical methods which require the repeated calculation of the time propagator for the system with time-dependent Hamiltonian[54]. In this paper we introduce a new method based on prediction theory[30] and nonlinear entrainment [42, 15, 17, 40], which makes it possible to derive the optimal field for the excitation of a di-atomic molecule or similar quantum systems in a closed form. These solutions provide physical and mathematical insights into the nature of optimal driving forces and can be quickly numerically calculated. As an instructive example, the method is applied to the stimulation of a Morse oscillator. The parameters were chosen to fit the vibrational modes of the HF molecule in order to compare to previous calculations for diatomic molecules[55]·[57]·[60]. It is shown that the method provides electric fields well below the ionization threshold[61], that lead to very high dissociation rates in comparably short time. Optimal driving forces which break the bonding of an HF-molecule within 1ps or faster turn out to be similar to chirped sine waves and are close to optimal driving forces of the corresponding classical oscillator[62]. In contrast, slow optimal excitations are of smaller amplitude and are composed of a series of monochromatic, bell shaped pulses which force the molecule to jump to higher and higher vibrational levels. In both cases, optimal driving forces turn out to be very smooth, i.e. all Fourier components which are larger than the basic frequency have a very small amplitude.

We study a quantum system like vibrations of a di-atomic molecule or an atom interacting with radiation that is described by the Hamiltonian

$$H(t) = \frac{1}{2m}p^2 + V(r) + E(t)er \tag{60}$$

where p is the momentum operator, m the mass, $E(t)$ the applied electric field, and e is the effective charge or dipole gradient. We assume that the wavelength of the applied electric radiation is large compared to the size of the atom or molecule.

In order to simplify the notation we set $\hbar, m$ and e equal to unity. An optimal time dependence is determined using a variational method which seeks to maximize the energy transfer to the quantum system given by $\Delta H = \int_0^T (d/dt < \psi|Ep|\psi >) dt$ while keeping the the energy fluence of the electric field $\bar{E}^2 = (1/T) \int_0^T E^2 dt$ fixed (strong resonance). Another constraint is given by the time dependent Schrödinger equation:

$$i \frac{\partial}{\partial t} \psi = H(t) \psi \tag{61}$$

These constraints lead to the condition that the variation of the functional $S = \int_0^T L dt$ has to be stationary[54], i.e.

$$\delta_{E, \dot{E}, \psi, \dot{\psi}} S = 0 \tag{62}$$

where

$$L = < \psi|E^2|\psi > + \kappa < \psi|Ep|\psi > + \frac{1}{2} Re(< \lambda||i\dot{\psi} >) \tag{63}$$

where $\lambda(t)$ is a Lagrange multiplier. κ may be considered as another Lagrange multiplier which fixes the average square value of the applied electric field $E(t)$ or as a weight in a penalty function[54]. L which balances the energy transfer against the size of the applied electric field. In any case, if κ is small the stimulation is fast and vice versa. From equation (62) the set of following Euler equations can be deduced:

$$E - \frac{1}{2} Re(< \lambda|r|\psi >) + \kappa < p > = 0 \tag{64}$$

$$|i \frac{\partial}{\partial t} \lambda > = H(t)|\lambda > - \kappa Ep|\psi > - E^2|\psi > \tag{65}$$

It can be shown[62] that

$$E(t) = \frac{2}{t + c} < p > \tag{66}$$

where c is a real constant, and $\lambda = i l_1(t) x \psi + i l_2(t) p \psi + i l_3(t) \psi$ where $l_1(t) = \kappa E(t) - \dot{l}_2$, $l_2(t) = (\frac{2}{t+c} + \kappa) < p >$, $l_3(t) = \int_0^t (E^2(\tilde{t}) - l_2(\tilde{t})) d\tilde{t}$ solves Eq.(64) for harmonic potentials and in the semiclassical approximation for any potential. Since the structure of the solution is very similar to optimal control forces of classical oscillators it might be possible to show that this solution or a very similar solution is correct for nonlinear potentials. In any case, $E(t) = \frac{2}{t+c} < p >$ satisfies the condition for weak resonance, as do all other driving forces of type $E(t) = f(t) < p >$, where $f(t) > 0$, since energy transfer $P = < Ep >$ is positive:

$$P = < Ep > = f(t) < p >^2 > 0 \tag{67}$$

The dynamics of the controlled system is given by a nonlinear, autonomous Schrödinger equation

$$i\dot{\psi} = (\frac{1}{2} p^2 + V(r) + E(t) r) \psi \tag{68}$$

where

$$E(t) = \frac{2}{t+c} <\psi|p|\psi> \tag{69}$$

In practice, one would like to control the system without feedback, whereas in Eq. 68 the applied field depends on the wave function ψ of the experimental system(Eq. 68). In order to achieve an open loop control we predict the response from a model

$$i\dot{\phi} = (\frac{1}{2}p^2 + \tilde{V}(r) + E(t)r)\phi \tag{70}$$

where

$$E(t) = \frac{2}{t+c} <\phi|p|\phi> \tag{71}$$

which represents the best theoretical knowledge of the dynamics of the controlled experimental system and which replaces Eq. 69 through Eq. 71. This means that we compute the optimal time dependence of the electric field through a simulation of the stimulation process and then apply the resulting optimal driving force to the experimental system. If the model is perfect, i.e. $\tilde{V} = V$, and $\phi(r,0) = \psi(r,0)$, then $\phi(r,t) = \psi(r,t)$ is valid for all times t.

Eq.66 and Eq.67 provide a physical interpretation of the optimal electric field: As can be seen from the above equation, the electric field is always parallel to the current. In addition, the electric field is in phase with the current $j = ex\psi$ all the time, despite the fact that the period of oscillation of the current changes as a function of time. Due to Eq.67, the reaction power is zero for optimal fields, i.e. the stimulation satisfies a weak resonance condition.

Numerical simulations [62] indicate that the energy transfer drops substantially if the model differs from the experimental dynamics. This feature can be used to improve the model through a systematic search in the parameter space. This procedure is called Nonlinear Quantum Resonance Spectroscopy(NQRS). It is a generalization of resonance spectroscopy. If the experimental system is initially in a stationary state and the model is therefore also chosen to be initially in the same stationary state the expectation value of the momentum $<\phi|p|\phi>$ is equal to zero. Therefore the system will never evolve to higher states. In the numerical calculation the excitation can be started by applying a small disturbance to the initial wave function of the model. An intriguing but unproven idea is that the the dynamics of the optimally stimulated system could be just the time reversal natural decay due to spontaneous emission of a highly stimulated system, in analogy to strong resonances of damped nonlinear oscillators.

The next problem we want to solve with the principl of the dynamical key also deal with wave phenomena: interference patterns. Andv the corresponding topic is the possibility of getting simple diffraction pattern from an amorphous material.

In the last few years a lot of models have been proposed for spatially chaotic or "turbulent" (RUELLE) [63] solid state configurations. There should be chaotic states in incommensurate crystals [64], and amorphous materials were suggested to be spatially chaotic configurations [65]. The existence of temporal chaotic phenomena has been verified experimentally, e. g. in the conductivity of BSN–crystals [66], but for spatial chaos there has never been direct evidence.

When investigating the structure of aperiodic (f. e. amorphous) materials one usually gets an unspecific diffraction pattern with very broad peaks [67]. This is even the case when the underlying algorithm for generating the arrangement of the scatterers is very simple. REICHERT and SCHILLING have shown [68] that diffraction patterns from a one dimensional chain of atoms are very similar to those of glasses if the separation of neighboring atoms is calculated by a map which is related to the Bernoulli shift. SADOC and MOSSERI proposed that the structure of an amorphous material could be generated by projecting a [3,3,5]–polytope in four dimensional space onto real three dimensional space [69] with a well defined deterministic algorithm. So there is the question about distinguishing between these deterministic structures and truly random ones.

The diffraction pattern from a monochromatic sine wave reflects the two–particle correlation function of the material, which is by definition a partition function of the distances of two atoms. The problem is that the averaging process inherent in the correlation function hides all the specific information about the simple algorithm which might be underlying the atomic structure. Different algorithms for the atomic arrangement can generate diffraction patterns which are quite similar.

This is mainly due to the fact that whenever monochromatic radiation is used, the time dependence of the incoming wave has no similarity whatsoever to the spatial structure of an aperiodic material. So one can ask about special effects when using incident waves with dynamics very similar to the structure of the material, i. e. when dynamics and structure are generated by the same algorithm.

A very similar problem occurs with resonant stimulation of damped, nonlinear oscillators. A general feature of nonlinear oscillators is the amplitude frequency interaction[45]. Due to the amplitude frequency interaction the dynamics of a damped nonlinear oscillator is aperiodic. If such an oscillator is stimulated sinusoidally the response is of small amplitude [46] and complicated [43]. Recently it has been shown that the response is simple and many orders of magnitude larger [42] if a special aperiodic driving force is used where the period of the perturbation and the eigenfrequency of the oscillator match at all amplitudes.

Following this line of argument, we investigate in this section the diffraction pattern from an aperiodic arrangement of scatterers using special incident waves. We present a certain class of incident waves which produce diffraction patterns with a small number of sharp peaks from an aperiodic structure. We show that it is possible to extract from these diffraction patterns simple deterministic rules providing the spatial structure. Finally, we discuss shortly possible experimental realizations of the proposed methods.

Fig.14 illustrates the arrangement we investigated. For simplicity we investigate one

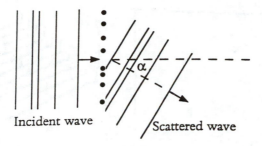

Incident wave   Scattered wave

Figure 14: *Sharp diffraction peaks from disordered structures. Coherent plane waves with a time dependent frequency are diffracted from an one dimensional aperiodic array of scatterers.*

dimensional structures in a two dimensional space. The scatterers are located at the positions $r^i = (r_1^i, r_2^i) = ((i + \delta l_i)L, 0)$ where $i = 1, 2, \ldots, n$ and where the variation of the separation between nearest neighbors δl_i is calculated by a map. We use two different maps. The map

$$\delta l_{i+1} = g\left(\delta l_i, \delta_L\right) := -sign\left(\delta l_i\right) \delta_L / 2 \tag{72}$$

produces a structure where the separation between nearest neighbors oscillates periodically. The parameter $0 < \delta_L < 1$ determines the magnitude of this oscillation. The other map

$$\delta l_{i+1} = f\left(\delta l_i, c_L\right) = c_L \delta l_i (1 - \delta l_i) \tag{73}$$

where $0 < c_L < 4$ and $0 < \delta l_i < 1$ produces structures where the separation between nearest neighbors oscillates aperiodically for certain values of the parameter c_L[6].

The incoming plane wave is $E_0\left(r_1, r_2, t\right) = E_0\left(r_2 - ct\right)$ where r_1, r_2 are the spatial coordinates, t is time and c is the speed of the wave. We use either sine waves $E_0 = \sin\left(r_2 - ct\right)$ or special waves defined by

$$E_0(t) = \begin{cases} \sin(2\pi t/\Delta) & \text{for } i + \delta t_i \leq t \leq i + \delta t_i + \Delta, \\ 0 & \text{else,} \end{cases} \tag{74}$$

where $\Delta \ll 1$ is a parameter and where the dynamics of δt_i is calculated by

$$\delta t_{i+1} = g\left(\delta t_i, \delta_t\right) := -sign\left(\delta t_i\right) \delta_t / 2 \tag{75}$$

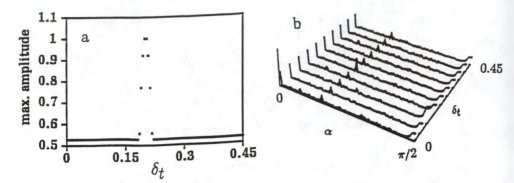

Figure 15: *The maximum amplitude* $\max_t |E(t,\alpha)|$ *versus* α *and* δt *, where* $\delta_L = 0.2$ *(b) and maximum amplitude of the diffraction peak* $\max_t |E(t), \alpha_0|$ *versus* $\delta t(a)$.

produces periodic plane waves where the separation between nearest maxima oscillates periodically, or

$$\delta t_{i+1} = f(\delta t_i, c_T) = c_T \delta t_i (1 - \delta t_i) \tag{76}$$

where $0 < c_T < 4$ and $0 < \delta t_i < 1$. The initial conditions for Eq.73 and Eq.76 are different but are close to the attractors. The parameter $0 < \delta_t < 1$ in Eq.75 determines the magnitude of frequency modulation.

The amplitude of the scattered wave, evaluated at position $r = (r_1, r_2)$ and time t, is computed as:

$$E(\vec{r}, t) = \sum_{i=1}^{n} f_i(\alpha) \frac{E_0(t - |r - r^i|/c - b_i \sin \alpha_0)}{|r - r^i|} \tag{77}$$

α the angle of observation. $f_i(\alpha)$ represents the angle–dependency of the amplitude factor of each scatterer. Here, we set $f_i(\alpha) = const = 1$ for convenience. In our calculations, we use $r = 10000$, $n = 21$ and $c = 1$.

We focus our attention to the signal emitted in the direction $\alpha_0 = \arcsin(tc/l)$. If $\delta_L = \delta_t$ all the scatterers will emit pulses with exactly the right phase difference for constructive interference. Hence the outgoing wave will have maximum amplitude. This is clearly shown in Fig.15 and in Fig.16. For periodic structures(Fig.15) the linewidth of the peak is $2\Delta/3$, because $\Delta/3$ is the largest phase difference between two sine waves producing a maximum of the wave larger than 1. Therefore, the line would be infinitely sharp when using δ–functions instead of sine waves.

Now let us look at our second example. Fig.16 shows that there is a sharp diffraction peak at $\alpha = \alpha_0$. The diffraction peak reaches a maximum amplitude when $c_T = c_L$. It

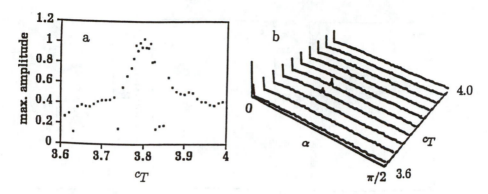

Figure 16: *The maximum amplitude* $\max_t |E(t, \alpha)|$ *versus* α *and* c_t *(b) and maximum amplitude of the diffraction peak* $\max_t |E(t), \alpha_0|$ *versus* c_T, *where* $c_L = 3.8$ *(a)*.

might be that the quality of this peak could be improved by choosing another function instead of the patchwork of sine oscillations in the incoming wave.

Experiments to confirm these numerical results could be done with water waves scattered by a grid of wires. Another possibility would be to use microwaves scattered by an arrangement of metal bars. A frequency and amplitude modulated laser perhaps could provide light waves useful for examining the effect on microgrids. Such grids were used e.g. by CHAIKIN et al. to examine the magnetic and superconducting properties of quasiperiodic networks [70]. Applications could be the checking of product quality of industrially manufactured microstructures. Here one would use an incoming wave reflecting the goal structure of the product. Any item not showing resonance could be omitted.

Ultimately, there is the question whether there are radiation sources which could emit waves of the above mentioned type, in the right frequency range to be useful for the structural analysis of solids. One possibility of such a source might be to use synchrotron radiation.

In summary, we have presented numerical simulations suggesting that specially shaped waves are suitable means to determine the parameters of an algorithm generating complex (especially chaotic) spatial structures. The first peak in the diffraction pattern is enhanced when the dynamics of the scattered wave and the spatial structure are generated by similar algorithms.

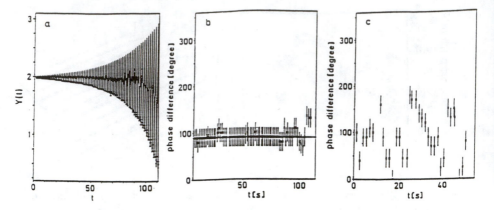

Figure 17: *(a) The extrema of the amplitude of the experimental pendulum• and the goal dynamics x (continuous line) versus time. (b) The phase difference φ between amplitude and driving force versus time for a resonant perturbation where $\eta_1 = 6 \ 10^{-4} kgm^2s^{-1}$, $\eta_2 = -.17\eta_1$. If the driving force is sinusoidal, i.e. $F = 4sin(\omega_0 t)$, where ω_0 is the basic frequency at the minimum of the potential the phase difference keeps not close to 90 degrees(c).*

4.3 Experimental Applications of the Principle of the Dynamical Key

As a real physical oscillator we used a damped wheel with an eccentric mass distribution[71]. The dynamics of the pendulum can be modeled by

$$\Theta\ddot{y} + \eta_1\dot{y} + c_1y + c_2sin(y) = F(t) \tag{78}$$

where y is the angular displacement, $\Theta = 1.65 \ 10^{-3}kgm^2$, $c_1 = 1.62 \ 10^{-2}kgm^2s^{-2}$, $c_2 = 0.03563kgm^2s^{-2}$ and η_1 is a friction constant which can be varied in the range $5 \ 10^{-5} kgm^2s^{-1} < \eta < 6 \ 10^{-4}kgm^2s^{-1}$. The time dependent driving force $F(t)$ is transmitted by a digital to force converter from a computer to the experimental pendulum.

Fig.17 shows a resonant stimulation of the pendulum, where the driving force was calculated by a time reflected Poincaré map [72]. Due to the strong amplitude-frequency coupling there is a sensitive dependence of the basic frequency of the oscillator from the amplitude of the oscillation. Fig.17b illustrates that the phase relation between the driving force and y remains 90 degrees despite the large shift of the frequency. The phase shift of 90 degrees indicates that the reaction power is zero and that the driving force is resonant[42]. If the driving force is sinusoidal there is no 90 degree phase difference between y and F. This experiment illustrates that a resonant stimulation of a low dimensional nonlinear mechanical oscillator is possible. Even this mechanical pendulum has in

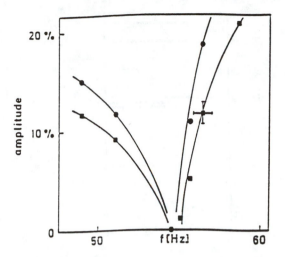

Figure 18: *The boundaries of the region of entrainment for sinusoidal (•) and for square wave perturbations(⊔) versus the basic frequency of the perturbation. The position of the hot wire probe is 6 mm behind the cylinder (∅ = 2mm) in the center of the vortex street(Reynold number = 70). The amplitude (distance between the extrema) of the acoustic perturbation is normalized to the amplitude of the variation of the velocity of the unperturbed system at this position.*

principle a large number of different vibrational states because it is a macroscopic object. But in general these mechanical pendula are constructed in such a way that the basic frequencies of all those vibrational modes are much higher than the basic frequency of the oscillation. The stimulation of vibrational modes can be kept small by an appropriate construction of the pendulum. Completely different is the situation if one tries to control the dynamics of a hydrodynamic system.

In hydrodynamic systems the basic frequency of slaved variables can not be shifted. If the dynamics of some order parameters of the system is reconstructed from an experiment [32, 30], the dynamics of the slaved variables is not known. However, a general feature of the response of a huge variety of oscillators is that the excitation is small if the perturbation is small. Therefore, resonant driving forces seem to be most appropriate in order to control hydrodynamic systems if just an order parameter equation is known. Recently it has been shown that the periodic dynamics of a velocity signal in the vortex street behind a circular cylinder can be modeled by a special low dimensional differential equation[35]. If the goal of a perturbation is to shift the basic frequency of the vortex street, stimulations with square waves are resonant perturbations of these special differential equations[73]. Fig.18 illustrates that the region of entrainment of the vortex street can be enlarged by using square waves. This indicates that the low dimensional model

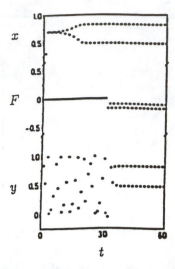

Figure 19: *Control of the chaotic dynamics of a logistic map without feedback, where $b =$ 3.2 (Eq.82). Illustrated are the goal dynamics x, the driving force F and the experimental dynamics y. The control force is off for $0 < t < 30$ and afterwards switched on (from [75]).*

can be used to predict the response of the high dimensional complex system, if the driving force is resonant.

4.4 Dynamical Control

The problem of finding nonlinear resonances is closely related to the problem of optimal control of nonlinear systems. Whereas extensive literature exists on the subject of linear controls with feedback and without feedback [74], little is known about their nonlinear counterparts [42, 15, 54]. Recently, progress in chaos theory [42, 30] has led us to a new approach to the control process [11, 17]. We assume that the dynamics of the order parameters of a complex system can be modeled by a differential equation of the type

$$\dot{x}(t, r_1, r_2, \ldots) = f\left(x_1, x_2, \ldots, t, r_1, r_2, \ldots, \frac{\partial}{\partial r_1}, \frac{\partial}{\partial r_2}, \ldots, F(t, r_1, r_2, \ldots) + p_m(t, r_1, r_2, \ldots)\right) \tag{79}$$

where y is a n-dimensional vector of order parameters and f is a n-dimensional flow vector field which depends on the order parameters $x_1, x_2, \ldots$, spatial coordinates $r_1, r_2, \ldots$, time t, and spatial derivatives. F is a driving force which acts on a set of significant control parameters p which is assumed to be close to the true parameter setting $p_m \approx p$. The goal of the control process is to make the limiting behavior of $x(t- > \infty)$ and a given goal dynamics $y(t)$ equal, by using an appropriate F.

Now we have to distinguish between two cases: (i) <u>dynamical control</u>, where no information about the actual state of the experimental system is required[11, 75], and (ii) feedback based controls [15]. For dynamical control complete entrainment occurs if the driving force makes the flow vectors of the experimental dynamics and the goal dynamics equal at the location of the goal dynamics, *i.e.*

$$\dot{y}\,(t, r_1, r_2, \ldots) = \\ f\left(y_1, y_2, \ldots, t, r_1, r_2, \ldots, \tfrac{\partial}{\partial r_1}, \tfrac{\partial}{\partial r_2}, \ldots, F\,(t, r_1, r_2, \ldots) + p\,(t, r_1, r_2, \ldots)\right) \tag{80}$$

and if the special solution $\mathbf{x}(t) = \mathbf{y}(t)$ is stable [13]. If $y(t)$ is not differentiable, Eq.80 must be analyzed in terms of distributions. Fig.19 illustrates the control of a chaotic motion with dynamical control. In this case the experimental dynamics is a chaotic logistic map dynamics:

$$x(t+1) = 4x(t)(1 - x(t)) + F(t) \tag{81}$$

and the goal dynamics is given by another logistic map dynamics

$$y(t+1) = 3/8 + z(t)/4 \tag{82}$$

$$z(t+1) = bz(t)(1 - z(t)) \tag{83}$$

where $0 < x(t) < 1$, $0 < z(t) < 1$, $0 < b < 4$, and $t = 1, 2, \ldots$. In order to apply Eq.79 and Eq.80 to that problem we have to replace $\dot{y}(t)$ by $y(t+1)$ to do a Taylor expansion of Eq.80. A comparison of coefficients in the expansion yields:

$$F(t) = y(t+1) - 4y(t)(1 - y(t)) \tag{84}$$

This driving force is independent of the actual states of the experimental system and can be calculated before the experiment is done. The stability of the control can be calculated as follows

$$\epsilon(t+1) = x(t+1) - y(t+1) \approx \left.\frac{\partial x(t+1)}{\partial x(t)}\right|_{x(t)=y(t)} \epsilon(t) \tag{85}$$

where $\epsilon(t) = x(t) - y(t)$ is the distance between goal and real dynamics. $\epsilon(t)$ tends to zero if

$$\lambda_c = \lim{}_{T \to \infty} \frac{1}{T} \sum_{t=1}^{T} \log \left| \frac{\partial x(t+1)}{\partial x(t)} \right|_{x(t)=y(t)} \Big| \le 0 \tag{86}$$

where λ_c is called control coefficient. For the logistic map in Eq.81 the control coefficient is smaller than zero for any goal dynamics $y(t)$ where $3/8 < y(t) < 5/8$. A generalization of this stability analysis for higher dimension systems and continuous systems is straightforward. The definition of the control coefficient is similar but not equal to the definition of Liapounov exponents

$$\lambda_x = \lim{}_{T \to \infty} \frac{1}{T} \sum_{t=1}^{T} \log \left| \frac{\partial x(t+1)}{\partial x(t)} \right|_{x(t)} \Big| \tag{87}$$

$$\lambda_y = \lim{}_{T \to \infty} \frac{1}{T} \sum_{t=1}^{T} \log \left| \frac{\partial y(t+1)}{\partial y(t)} \right|_{y(t)} \Big| \tag{88}$$

where λ_x and λ_y are the Liapounov exponents for the unperturbed experimental system and the goal dynamics respectively. This means that the goal dynamics and the experimental dynamics can both be chaotic, i.e. $\lambda_x > 0$ and $\lambda_y > 0$, and nevertheless the control is stable since $\lambda_c < 0$. This happens, for example, for $b = 4$ in the example given above.

For feedback controls[15] the flow vectors of the experimental dynamics and the goal dynamics have to be made equal, too, and therefore they are very similar to dynamical control. However, since information about the actual state of the system can be used one has

$$
\begin{aligned}
&\dot{y}\,(t, r_1, r_2, \ldots) = \\
&f\left(x_1, x_2, \ldots, t, r_1, r_2, \ldots, \tfrac{\partial}{\partial r_1}, \tfrac{\partial}{\partial r_2}, \ldots, F\,(t, r_1, r_2, \ldots) + p\,(t, r_1, r_2, \ldots)\right)
\end{aligned}
\tag{89}
$$

the special solution $x(t) = y(t)$ is usually stable. This is a distinct advantage of feedback controls over dynamical controls. Some targets are not reachable with dynamical controls since they are unstable. It is not possible to get a stable dynamical control if the goal dynamics has a stable limit cycle at a location where the experimental system has an unstable limit cycle, whereas feedback controls can deal with this situation[15]. However, feedback controls need the information about the actual state of the system, which in many physical systems, as for example quantum systems, is not accessible. Any delay between the measurement and the resulting adjustment of the control force causes the instabilities. All feedback controls need continuous access to sensors and a control unit which has a typical time scale much faster than the typical time scale of the experiment, whereas dynamical control needs no feedback after an initial modeling period where an estimate of p is determined, and no feedback at all, if a model of the experimental dynamics is known. In addition, the time scale of the control unit does not need to be faster than the typical time scale of the experiment and is in most applications much slower. For example, in order to control the motion of a hot nucleus, a spatial aperiodic sequence of magnetic and electric fields which the nucleus is traveling through, could be preassigned. In this case the time constant of the control unit is zero. Since the control force is calculated for both types of control in the same way, the size of the control force is the same.

Therefore, in the following we focus on dynamical control. As soon as the goal dynamics can be integrated in a closed form, one also has a closed form for the response of the driven experimental system, even if the driving force is aperiodic.

A driving force is called a resonant perturbation if it forces the experimental system to satisfy a certain set of constraints like, for example, to transfer a certain energy or to reach a certain target and be extraordinary small. In this case, the variational principle establishes a link between the goal dynamics and the experimental dynamics, which is the similarity mentioned in the principle of the dynamical key. The type of similarity which is established between the two dynamics depends on the constraints. In general, the goal equation is equivalent to the equation of motion resulting from the variation. In addition, in some cases solvability conditions of the variational calculus determine reachable targets.

4.5 General Resonance Spectroscopy

In section 3 we presented a method to extract a model from a given time series: dynamical reconstruction. Dynamical reconstruction provides accurate models for the natural dynamics of the system. However, these models are only useful for a control if there is an overlap between the natural dynamics of the system and the goal dynamics. In particular, if the experimental dynamics is simple, for example a fixed point dynamics, this is not the case. In this case it is necessary to explore the state space through appropriate perturbation. These perturbations have to be small in order to avoid the stimulation of slaved variables, as we argued earlier. A systematic procedure for such an active modeling is general resonance spectroscopy.

In linear resonance spectroscopy, one applies a sinusoidal driving force to a harmonic oscillator. Such an oscillator has only one parameter, the characteristic frequency. The amplitude of the resulting stimulated oscillator is a function of the driving frequency and a maximum when the driving frequency is equal to the oscillator's characteristic frequency. One determines the characteristic frequency of the oscillator, the model parameter, by systematically varying the driving force until this resonance, i.e. large response is observed.

In other words, one is seeking to match two quantities - the driving frequency and the characteristic frequency - using the amplitude of the oscillation as a measure of how good the match is.

This procedure can be generalized. The central concept is simple: If our model of some experimental dynamics is a good one, then we should be able to predict the response of the driven experimental system. Using the deviation $Q = \overline{|z(t) - x(t)|}$ between the prediction $z(t)$ and the real trajectory of the driven system $x(t)$, we find how good the model is. By varying the parameters in the model to minimize this deviation, we find the "general resonance" and thus fine-tune the model. $1/Q$ becomes infinitely large at this resonance and indeed CHANG et al. have shown that for harmonic oscillators $1/Q$ as a function of the characteristic frequency is closely related to standard resonance curves.

The goal dynamics must be chosen with some care in order to make the modeling process successful. The control must be stable for a perfect model, so that small perturbations in the initial conditions remain small. Also, for modeling purposes, one wishes the goal dynamics to cover a large area of the state space, as the model derived by this procedure is valid only over the range of the goal dynamics. For the same reason, one would prefer a chaotic goal dynamics to a periodic one.

Finally, for experimental considerations, one would like (i) to keep the driving force as small as possible to avoid exciting modes that one is not interested in and (ii) a goal dynamics which is significantly different from the natural dynamics in terms of the

46

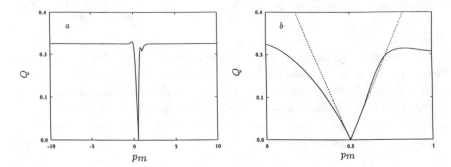

Figure 20: *A general resonance curve[12, 49]. Plotted is Q versus p_m for Eq.97 and Eq.98, where $p = 0.5$ (a). (b) shows the same resonance curve on a different scale (— numerical result, - - – theoretical result from Eq.96).*

experimental noise level, since in most experimental systems Q can be accurately measured rather than $1/Q$ we investigate the resonance curve determined by Q. In this case a rough estimate of the signal to noise ratio of the resonance curve is given by $Q = \overline{\|z(t)\|} - \overline{\|x(t)\|}/\sigma_D$ where σ_D is the standard deviation of the noise level.

As an example we consider discrete one dimensional maps:

$$x(t+1) = f_E(x(t)) + F^{noise}(t) + F(t) \tag{90}$$

where F is a time dependent driving force. Since we assume the exact form of the experimental map to be unknown, a model

$$y(t+1) = f_M(y(t)) + F(t) \tag{91}$$

where p_m is a set of model parameters, is used for the control. To achieve the desired goal dynamics, z(t), the driving force should be

$$F(t) = -f_M(z(t)) + z(t+1) \tag{92}$$

If we expand f_E and f_M in a space of orthogonal functions, for example a Fourier expansion, Eq.90 and Eq.91 can be rewritten, at least formally, as

$$x(t+1) = f_E(x(t)) + F(t) = f(x(t), p) + F^{noise}(t) + F(t) \tag{93}$$
$$y(t+1) = f_M(y(t)) = f(y(t), p_m) \tag{94}$$

where p and p^m represent a set of expansion coefficients for f_E and f_M, respectively.

Taylor expanding f around $z(t)$ and p_m, and keeping the first order terms we get for $\epsilon(t) := x(t) - z(t)$:

$$\epsilon(t+1) = \left.\frac{\partial f(x,c)}{\partial x}\right|_{z(t),p^m}\epsilon(t) + \sum_j \left.\frac{\partial f(x,c)}{\partial p_j^m}\right|_{z(t),p^m}(p_j^m - p_j) \tag{95}$$

For large T the mean square separation between goal dynamics and the actual dynamics is

$$Q = 1/T\sum_{t=1}^{T}\epsilon^2(t) \approx \sum_j \kappa_j(p_j^m - p_j)^2 \tag{96}$$

where $\kappa = 1/T\sum_{t=1}^{T}\left.\frac{\partial f(x,c)}{\partial p_j^m}\right|_{z(t),p^m}$ if the goal dynamics is chosen in such a way that the control is very stable for a correct model, i.e. $\left.\frac{\partial f(x,c)}{\partial x}\right|_{z(t),p^m} \approx 0$.

As an example we use for "unknown real system" a Gaussian map:

$$x(t+1) = 1.6p\exp\left(-8(x(t)-p)^2\right) + F^{noise}(t) \tag{97}$$

for the model:

$$y(t+1) = 1.6p_m\exp\left(-8(y(t)-p_m)^2\right) \tag{98}$$

where p_m is an unknown model parameter. The goal dynamics z is uncorrelated band limited white noise, where $0.25 < z(t) < 0.75$. $F^{noise}(t)$ is uncorrelated, noise with standard deviation $\sigma = 0.0001$. Fig.20 shows the corresponding resonance curve.

5. PHYSICAL INTERPRETATION OF AIMS

In the previous sections we have discussed methods to extract AIMs and the dynamics on the AIM and illustrated how to use them for a control. We illustrated that the geometrical shape of the AIM can be complicated, and that the flow on the AIM tends to be highly nonlinear. Therefore, an important question arises: Is it possible to attach any physical meaning to those AIMs? Is it possible to extract them from variational principles? What are the corresponding conserved quantities? Well, the motion of macroscopic mechanical objects describes the flow on those AIMs since all internal vibrational modes are neglected, and it is well connected to physical principles, as, for example, Hamilton's principle and the conservation of energy. Solitons and the associated conservation laws are other examples. But how about more complicated, dissipative systems? Is there a variational principle for turbulent states? A step in this direction may be the following two paradigms.

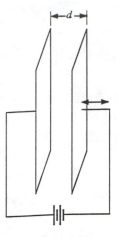

Figure 21: *A capacitor made of a movable plate and a fixed plate is charged by a battery. If no external potential forces, as, for example, gravity act on the movable plate, then the movable plate approaches the fixed plate. The speed of the motion is determined by friction, inertia, and the amount of charge on the plates.*

5.1 The Principle of Minimum Resistance, Regenerative Materials

The principle of <u>minimum resistance</u> states that *in a strong nonequilibrium system the state with the smallest resistance is stable, where the resistance may be an electrical resistance, thermal resistance etc.* Since there is no general proof yet, we are presenting the general idea in two examples. This principle might have important application: If the operating state of an industrial product is the state with the smallest resistivity, then it returns to that state after a minor damage, i.e, it is a regenerative system, as for example the self-repairing wire in Fig.1.

The first system is a macroscopic electric capacitor consisting of two metallic plates (Fig.21). We assume that the gap between the plates is filled with a dielectric fluid with small but nonzero ohmic conductivity and that the resistance R between the two plates is directly proportional to the gap size d, i.e. $R \sim d$. Further we assume that one of the plates can move, i.e. change the width of the gap, and that potential forces, as, for example, gravitational forces, act on the movable plate. As soon as a battery with constant voltage or any other current source is applied, the movable plate approaches the other plate and the gap size monotonically decreases. Inertial forces, friction forces and the size of the applied electrical voltage determine the velocity of the movable plate, but the limiting state is in any case a zero gap size. If the current I through the plates is kept constant, the entropy production due to the ohmic resistivity and the energy content of the electric field between the plates decrease monotonically. In the limit of zero gap both are zero, i.e. as small as possible as predicted by PRIGOGINE's maximum entropy production[76] principle. However, if a constant voltage V is applied, both entropy production and the energy content of the electric field increase monotonically. For current supply with a nontrivial $I(V)$ relation there is no general trend for entropy production or energy content. However, in all cases the resistance between the plates decreases.

The crucial point in this experiment is that (i) the forces are due to the gradient of energy densities and (ii) all those energy densities are associated with the flow, i.e. the nonequilibrium. Since the nonequilibrium system is an open system with a unlimited supply of energy, at least in principle, the total energy content is of minor importance for the limiting states, in contrast to closed systems. The basic idea is that any extra hindrance for the flow produces a gradient of the energy density (voltage gradient ∇V) according to Ohm's law ($\nabla V = IR/d$), Fourier's law or similar laws. This pressure gradient (=gradient of the energy density) produces a force F which acts on the hindrance and moves it if no other potential forces act on the hindrance. Where will the hindrance reach a stable stationary state? If at all, at a location where it produces no gradients of the energy density, i.e. in general out of the way of the flow. For the example with the capacitor, the hindrance is the dielectric liquid. It moves out of the way of the current, i.e. leaves the spacing between the two plates, causing the spacing between the two plates to collapse, and moves to a location where it does not interfere with the flow. This means the resistivity gets as small as possible, since the flow removes all hindrances, i.e. $\partial F/\partial d = 0$ at stationary states. Since $\partial F/\partial d = \partial F/\partial R * \partial R/\partial d$, $\partial R/\partial d = 0$ at stationary states if $\partial F/\partial R$ is not equal to zero for large t. Due to Ohm's law $\partial F/\partial R$ is not equal to zero for the capacitor and therefore, the stationary state is the state with minimum resistance, i.e. $\partial R/\partial d = 0$. If potential forces are present which are not caused by the flow, as, for example, a gravitational force acting on the movable plate, they may trap a hindrance in the flow and stabilize a state which is not of smallest resistivity. However, if potential forces are small compared to pressure gradients due to the flow, i.e. in strong external nonequilibrium, the state of smallest resistivity may be a good estimate for the final state.

Whereas the first example has more of an explanatory character, the second might have important applications. At the beginning of the experiment, N smooth steel balls (radius $r = 1mm$, mass $m = 33.0 \pm 1.3mg$) are homogeneously distributed at the ground of a horizontal cylindrical cell. The cylindrical cell (radius $r_{cell} = 57mm$)is made up by a thin layer of oil (height $h = 3mm$) within an acrylic dish and a layer of air above. The inner perimeter of the dish is equipped with a grounded metallic ring(Fig.22). From a metallic tip which is $40mm$ above the center of the dish, electric charges are sprayed upon the oil-surface. The difference of the electric potential V between a metallic tip and the grounded ring can be adjusted from $V = 10kV$ to $V = 25kV$, which gives a current I_0 on the order of $0.1\mu A$. Due to the electric current the metallic balls agglomerate. These agglomerates are photographed and digitized with a resolution of 14000 x 14000 pixels.

The working fluid is castor oil, chosen because it has a small conductivity ($\sigma_{oil}(20°C) \leq$

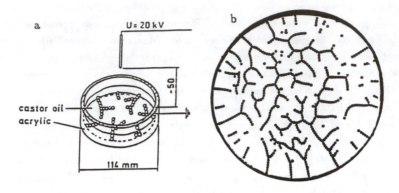

Figure 22: *Experimental set up(a) and a typical structure out of 600 balls at $V = 20kV$ (b). Systematic studies have been done by B. MERTÉ et al.[77].*

$10^{-12}(\Omega \cdot cm)^{-1}$), a high dielectric content ($\epsilon_{oil} = 6$), and a high viscosity ($\eta(20°C) = 0.990 \pm 1\% Pa \cdot s$). Because of the high dielectric constant and the small conductivity a lot of electrostatic energy can be stored in the oil while the heating of the fluid due to the Ohmic resistance is small. Due to high viscosity of the oil, the motion of the balls is slow compared to the relaxation time of the distribution of the electric charges.

In order to estimate the overall resistivity of the cell numerically, the following simple model is used. All metallic balls which are in contact with the metallic ring electrode are at zero voltage. We assume that the medium in between is an isotropic, permeable, conducting dielectric and that magnetic effects are neglectable. In this case the stationary state of the electrostatic potential ϕ results from a two-dimensional Poisson equation, $\Delta\phi(r) = S_t(t)S_r(r)/(\sigma_{oil} \cdot h)$ where r is the position and where S_t and S_r describe the time dependence and normalized spatial distribution ($\int S_r d\mathbf{x} = 1$) of the rate at which the charges are transferred to the two dimensional surface. We assume that in the experiment the electric loads are homogeneously spread from the counter electrode onto this surface at a constant rate, i.e. $S_t = I_0, S_r = 1/a$, where the area of the cell $a = r_{cell}^2 \pi$. With a standard relaxation algorithm the Poisson equation is numerically solved on a rectangular grid scheme (i, j) of $N_g^2 = 250000$ grid points which just covers the cell. The overall resistance is estimated by $R = h\sigma_{oil} \sum_{i,j}^{N_g, N_g}((\phi_{i+1,j}^s - \phi_{i,j}^s)^2 + (\phi_{i,j+1}^s - \phi_{i,j}^s)^2)/I_0^2$ where $\phi_{i,j}^s$ is the stationary state of the electric potential at the grid point (i, j) for a given distribution of balls. ϕ is set to zero at all locations which are outside the perimeter of the cell and at the locations of grounded balls. Balls which touch the electrode or touch another grounded ball are called grounded. Further we counted the number of grounded balls N and the number of tips M of the agglomerate. Tips are those grounded balls which touch just one other ball.

It can be experimentally demonstrated that the agglomerates converge to a stable stationary state: stationary structures return to their original state after a small external perturbation. Further, all the stationary patterns have been without any closed loop.

For a theoretical description we use 2-dimensional models which describe the horizontal motion of the electric charges and the balls. The balls are assumed to be ideal conductors, i.e. their resistivity $1/\sigma_{ball} \approx 0$, whereas the dielectric constant ϵ_{ball} has some limited value. The electric carriers are homogeneously injected into the oil at a rate which may depend on time. The boundary of the cell is assumed to be an ideal conductor at a fixed electrostatic potential. The dynamics of the electrostatic potential $\phi(r,t)$, the current density $i(r,t)$, and the density $\rho(r,t)$ of free carriers is given by the continuity equation $\partial\rho/\partial t = -j(r,t)+ S$, where $j = \nabla_r i$ is called the absorbed current, Ohms law $i/(h\sigma(r,q_i)) = -\nabla_r\phi$ and Poissons's law $\nabla_r h\epsilon(r,q_i)\nabla_r\phi = -\rho$. $\epsilon(r,q_i)$ and $\sigma(r,q_i)$ depend on the location of the balls and equal either ϵ_{oil} and σ_{oil} or ϵ_{ball} and σ_{ball}. The motion of the balls is modeled by

$$m\ddot{q}_i + \gamma\dot{q}_i = \nabla_{q_i}(W(q_i,t) + W_B) \qquad i = 1,2,\ldots,N \tag{99}$$

where q_i is the position of ball # i, γ is an effective friction constant,

$$W = \int_a h\epsilon(r,q_i)(\nabla_r\phi)^2 dr = \int_b h\epsilon_{oil}(\nabla_r\phi)^2 dr \tag{100}$$

is the electrostatic energy, where W_B is the work stored in the power supply. b is the area of the cell, excluding the balls. An analysis of dimensions yields estimates for the typical time scale of the motion of the electric carriers $t_e = \epsilon_{oil}/\sigma_{oil} \approx 1min$ and the motion of the balls $t_b = \gamma(\sigma_{oil}h)^2/(\epsilon_{oil}\bar{S}^2 v) \approx 30min$ for $\sigma_{oil} = 10^{-12}(\Omega \cdot m)^{-1}$, $\gamma \approx \gamma_{Stoke} = 6\pi r\eta$ and $\bar{S} = I_0/A$. $\bar{S}$ should be a typical value for $S(r,t)$, for example, it's maximum or it's mean value. $v = 4/3\pi r^3$ is the volume of the balls. Since t_e is much smaller than t_b we eliminate the fast variable ρ adiabatically [1], i.e. we assume $\rho(r,q_i,\dot{q}_i,\ldots,t) \approx \rho(r,q_i,t)$. Since the relaxation time of the carriers within the spheres is approximately zero we can separate time dependence caused by $S_t(t)$ from the space dependence in the whole system. In this approximation we get from the continuity equation and Ohms law for the motion of the carriers within the oil $\rho_x = S_r = j_x = 1/a, \nabla_r^2\phi_x = S_r$, $h\epsilon_{oil}\phi_t = \rho_t = \epsilon_{oil}j_t/\sigma_{oil}$, and $\dot{\rho}_t = -j_t + S_t$, where $j = j_t(t)j_x(r)$, $\rho = \rho_t(t)\rho_x(r)$, and $\phi(\mathbf{x},t) = \phi_t(t)\phi_x(r)$. Therefore, the overall resistance

$$R(q_i) := P_\Omega/j_t^2 = \frac{\sigma_{oil}}{\epsilon_{oil}}W/j_t^2 = \frac{1}{h\sigma_{oil}}\int_B(\nabla_r\phi_x)^2 dr \tag{101}$$

is uniquely determined by the positions of the balls $q_i(t)$ and is independent from S_t. P_Ω is the energy dissipation due to the ohmic resistance. Any ball within the liquid would be charged and therefore, attracted by the grounded balls. Therefore, we assume that all balls are grounded. In this case the entire charge within the oil $Q_{oils} := \int_b \rho dr = \int_b S(t)dr = S_t b/a$ equals the amount of charge induced at the spheres and at the boundary of the cell and is independent from q_i. The work stored in the power supply can not be stored in the cell through a movement of q_i and we have $\nabla_{q_i} W_B = 0$.

If the flux is not time dependent, i.e. $S_t = I_0$, $L = m \sum_{i=1}^{N} \dot{q}_i^2/2 + W$ is a Liapounov function [1] for the dynamics given by Eq. 99 the balls approach a stable stationary state where electrostatic energy W reaches a local or a global minimum. At stable stationary states the overall resistance $R = \sigma_{oil}/\epsilon_{oil} W/I_0^2$, as well as the energy dissipation [76] $P := I_0^2 R + \sum_{i=1}^{N} \gamma \dot{q}_i^2$ reaches a minimum, too. Since the flux is not space dependent, i.e. $S_r = 1/a$, we find, using Green's first theorem and Poisson's equation for the mean value of the electrostatic potential, $\bar{\phi}(q_i, t) := \int_B \phi dr/A = R(q_i)S_t(t)$. Therefore, if both S_t and S_r are constant, $\bar{\phi}$ is at stable stationary state at a minimum.

If $m\ddot{q}_i$ is negligibly small in Eq. 99, R is a Liapounov function for the motion of the balls if we rescale time like $\tilde{t} = \int S_t(t)^2 dt$. Since $d\tilde{t}/dt \geq 0$, R is monotonically decreasing in time. Since $\gamma_{Stokes} t_b/m \approx 10^3$ for the experiment, the dynamics is assumed to be overdamped, and we expect that R monotonically decreases while approaching a stable stationary state. This is in agreement with the experimental observations. If $m\ddot{q}_i$ is negligibly in Eq. 99, the trajectory of the balls $q_i(t)$ is independent from S_t. Therefore, all geometrical properties of the agglomerates, including the limiting value of R for $t- > \infty$, should be independent of the applied flux S_t. This is not completely consistent with our experimental findings. We assume, that due to a static friction force which has been neglected in Eq. 99, the dynamics does not reach exactly optimal states, but as soon as the flux is increased, it comes closer. This interpretation is in agreement with the observation that R decreases as a function of time and if S_t is increased.

We conclude that as soon as static friction is small the nonequilibrium structure approaches a stationary state where resistance $R(q_i)$ reaches a minimum, if the imposed flux S is not time dependent or if inertial term $m\ddot{q}_i$ is negligibly small. In both cases the dynamics of the balls tend to minimize the electrostatic energy W. The striking fact is, that the entire energy content the energy dissipation, and the space average of the potential may have a complicated time dependence, in contrast to the overall resistance.

The principle of minimum resistance might be important for other dendritic structures, as, for example, crystallization in a undercooled melt.

5.2 Principle of Optimal Interaction

The principle of <u>optimal interaction</u> states that *a state is stationary if the interaction between two given systems, for example, the exchange of energy or the exchange of information, is extremal*. In some cases it is stable, and in some cases unstable. In any case, if an experimental system can reach such a state then the likelihood of finding it in such a state is high, since it is stationary. The principle of minimum resistance might be a special case of this principle, but there is no proof yet. We will prove the theorem on optimal interaction for two conservative, nonlinear oscillators with a weak harmonic coupling[46].

We investigate Hamiltonians of the form

$$H = V_1(x_1) + \frac{1}{2}m_1\dot{x}_1^2 + V_2(x_2) + \frac{1}{2}m_2\dot{x}_2^2 + \frac{1}{2}k(x_1 - x_2)^2 \qquad (102)$$

where x_1 and x_2 are amplitudes, m_1 and m_2 are inertial constants, V_1 and V_2 are harmonic or anharmonic potentials, and k is a small constant which quantifies the magnitude of the coupling between the two oscillators. In order to separate the only term which depends on the amplitudes of both oscillators we rewrite Eq. (102)

$$H = H_1 + H_2 + kH_c \qquad (103)$$

where $H_1 := V_1 + \frac{1}{2}m_1\dot{x}_1^2 + \frac{1}{2}kx_1^2$, $H_2 := V_2 + \frac{1}{2}m_2\dot{x}_2^2 + \frac{1}{2}kx_2^2$, $H_c := -x_1x_2$ and kH_c is assumed to be a small perturbation of the integrable Hamiltonian $H_1 + H_2$.

Since the method of secular perturbation theory is well known and was applied to diverse problems [78], here only the main results are recalled, in order to illustrate the differences between the approach in the present paper and the usual method [79].

Since the system is integrable for $k = 0$, $H_1 = H_1(I_1)$ and $H_2 = H_2(I_2)$ are rewritten in their action-angle form where (I_1, Θ_1) and (I_2, Θ_2) are the corresponding action-angle variables for $k = 0$. The perturbation H_c is expanded in a Fourier series

$$\begin{aligned} H_c &= -\left(\frac{c_{1,0}(I_1; \gamma_1)}{2} + \sum_{l=1}^{\infty} c_{1,l}(I_1; \gamma_1)\cos(l\Theta_1)\right) \\ &\quad \cdot \left(\frac{c_{2,0}(I_2; \gamma_2)}{2} + \sum_{m=1}^{\infty} c_{2,m}(I_2; \gamma_2)\cos(m\Theta_2)\right) \end{aligned} \qquad (104)$$

where γ_n denotes the two sets of parameters of the potential and the inertial constant. $c_{1,l}, c_{2,l}$ are the Fourier components of the unperturbed oscillators

$$c_{n,l}(I_n; \gamma_n) = \frac{1}{\pi}\int_{\Theta_n^0}^{\Theta_n^0 + 2\pi} x_n(I_n, \Theta_n; \gamma_n)\cos(l\Theta_n)d\Theta_n \qquad (105)$$

where Θ_n^0 has to be chosen appropriately so that the Fourier series consists only of cosine-terms.

For a large variety of oscillators the basic Fourier component is considerably larger than all other Fourier components. Therefore, we introduce the following notation:

$$x_n(\Theta_n) = \frac{c_{n,0}}{2} + c_{n,1}\left(\cos\Theta_n + \sum_{l=2}^{\infty} \alpha_{n,l}\cos(l\Theta_n)\right) \qquad (106)$$

where $\alpha_{n,l} = \frac{c_{n,l}(I_n;\gamma_n)}{c_{n,1}(I_n;\gamma_n)}$. If higher Fourier components fall off rapidly $\alpha_{n,l} \ll 1$. For homothetic oscillators $\alpha_{n,l}$ is independent of the amplitude of the oscillation, i.e.

$$\frac{\partial \alpha_{n,l}}{\partial I_n} = 0 \qquad (107)$$

In the following we want to investigate the dynamics close to the primary resonance. In order to avoid secular terms we use the following action-angle variables [79]

$$
\begin{aligned}
\tilde{I}_1 &:= I_1 \\
\tilde{I}_2 &:= I_2 + I_1 \\
\tilde{\Theta}_1 &:= \Theta_1 - \Theta_2 \\
\tilde{\Theta}_2 &:= \Theta_2
\end{aligned}
\tag{108}
$$

For weak coupling the phase-difference $\tilde{\Theta}_1$ is changing more slowly than $\tilde{\Theta}_2$ and we can average [80] [81] over the fast variable $\tilde{\Theta}_2$

$$
\bar{H} = \frac{1}{2\pi} \int_0^{2\pi} H \, d\tilde{\Theta}_2 = H_1\left(\tilde{I}_1; \gamma_1\right) + H_2\left(\tilde{I}_2^0 - \tilde{I}_1; \gamma_2\right) - \tag{109}
$$
$$
-\frac{k}{4} c_{1,0}\left(\tilde{I}_1; \gamma_1\right) c_{2,0}\left(\tilde{I}_2^0 - \tilde{I}_1; \gamma_2\right) - \frac{k}{2} \sum_{l=1}^{\infty} c_{1,l}\left(\tilde{I}_1; \gamma_1\right) c_{2,l}\left(\tilde{I}_2^0 - \tilde{I}_1; \gamma_2\right) \cos\left(l\tilde{\Theta}_1\right).
$$

Since $\bar{H} = \bar{H}\left(I_1, I_2, \tilde{\Theta}_1; \gamma_1, \gamma_2\right) = \bar{H}\left(\tilde{I}_1, \tilde{I}_2 - \tilde{I}_1, \tilde{\Theta}_1; \gamma_1, \gamma_2\right)$ is independent of $\tilde{\Theta}_2$, $\tilde{I}_2$ is a conserved quantity in this approximation, i.e. $\tilde{I}_2(t) = \tilde{I}_2^0$, where $\tilde{I}_n^0 := \tilde{I}_n(t = 0)$.

Using the notation of Eq. (106), Eq. (109) reads:

$$
\bar{H} = H_1 + H_2 - \frac{k}{4} c_{1,0} c_{2,0} - \frac{k}{2} c_{1,1} c_{2,1} U\left(\tilde{\Theta}_1\right)
\tag{110}
$$

where

$$
\begin{aligned}
U\left(\tilde{\Theta}_1\right) &= U\left(\tilde{\Theta}_1, \tilde{I}_1, \tilde{I}_2^0 - \tilde{I}_1; \gamma_1, \gamma_2\right) \\
&:= \cos\tilde{\Theta}_1 + \sum_{l=2}^{\infty} \alpha_{1,l}\left(\tilde{I}_1; \gamma_1\right) \alpha_{2,l}\left(\tilde{I}_2^0 - \tilde{I}_1; \gamma_2\right) \cos\left(l\tilde{\Theta}_1\right)
\end{aligned}
\tag{111}
$$

If the dynamics of both oscillators are homothetic U is independent of $\tilde{I}_1$ and $\tilde{I}_2$.

The dynamics described by the Hamiltonian of Eq.(110) has several fixed points, three of them located at $\tilde{\Theta}_1^f = 0, \pm\pi$. The corresponding action $\tilde{I}_1 = \tilde{I}_1^f\left(\tilde{I}_2^0; \gamma_1, \gamma_2\right)$ can be calculated by

$$
\left.\frac{\partial\left(H_1\left(\tilde{I}_1\right) + H_2\left(\tilde{I}_2^0 - \tilde{I}_1\right)\right)}{\partial\tilde{I}_1}\right|_{\tilde{I}_1^f} - \left.\frac{k}{4}\frac{\partial}{\partial\tilde{I}_1}\left(c_{1,0}\left(\tilde{I}_1\right) c_{2,0}\left(\tilde{I}_2^0 - \tilde{I}_1\right)\right)\right|_{\tilde{I}_1^f} -
$$
$$
- \left.\frac{k}{2}\frac{\partial}{\partial\tilde{I}_1}\left(c_{1,1}\left(\tilde{I}_1\right) c_{2,1}\left(\tilde{I}_2^0 - \tilde{I}_1\right) U\left(\tilde{\Theta}_1^f, \tilde{I}_1, \tilde{I}_2^0 - \tilde{I}_1\right)\right)\right|_{\tilde{I}_1^f} = 0
\tag{112}
$$

The system is said to be at resonance if

$$
\omega_n^0 = \omega^f
\tag{113}
$$

where $\omega_n = \frac{\partial \bar{H}}{\partial I_n}$, $\omega_n^0 := \omega_n \left(\tilde{I}_1^0, \tilde{I}_2^0 - \tilde{I}_1^0 \right)$ and $\omega_1^f = \omega_2^f = \omega^f := \omega_n \left(\tilde{I}_1^f, \tilde{I}_2^0 - \tilde{I}_1^f \right)$.

Now we assume that the system is at time $t = 0$ close to resonance, i.e. $\tilde{I}_1^0 \approx \tilde{I}_1^f$, and we further assume that the transfer of action due to the coupling of the oscillators remains small, i.e. $\left| \tilde{I}_1(t) - \tilde{I}_1^f \right| \ll \left| \tilde{I}_1^f \right|$. Therefore we can expand $\bar{H}$ about $\tilde{I}_1^f$. Introducing $\Delta I := \tilde{I}_1 - \tilde{I}_1^f$ we have

$$
\begin{aligned}
\bar{H} &= \frac{1}{2} \left. \frac{\partial^2 \bar{H}}{\left(\partial \tilde{I}_1 \right)^2} \right|_{\tilde{I}_1^f} (\Delta I)^2 - \frac{k}{2} c_{1,1}^f c_{2,1}^f U^f \left(\tilde{\Theta}_1 \right) + \\
&\quad + const + O \left((\Delta I)^3, \Delta I k \right) \\
&= \frac{1}{2} M^f (\Delta I)^2 - \frac{k}{2} c_{1,1}^f c_{2,1}^f U^f \left(\tilde{\Theta}_1 \right) + const + O \left((\Delta I)^3, \Delta I k \right)
\end{aligned}
\tag{114}
$$

where $c_{n,1}^f := c_{n,1} \left(\tilde{I}_1^f, \tilde{I}_2^0 - \tilde{I}_1^f \right)$, $U^f \left(\tilde{\Theta}_1 \right) = U \left(\tilde{\Theta}_1, \tilde{I}_1^f, \tilde{I}_2^0 - \tilde{I}_1^f \right)$ and $M^f = M \left(\tilde{I}_1^f \right)$ measures the amplitude-frequency coupling of the whole system. M reads

$$
\begin{aligned}
M \left(\tilde{I}_1 \right) &= \omega_1 \frac{\partial \omega_1}{\partial \bar{H}} + \omega_2 \frac{\partial \omega_2}{\partial \bar{H}} \\
&= \Omega_1 \frac{\partial \Omega_1}{\partial H_1} + \Omega_2 \frac{\partial \Omega_2}{\partial H_2} - \frac{k}{4} \frac{\partial^2 c_{1,0} c_{2,0}}{\left(\partial \tilde{I}_1 \right)^2} - \frac{k}{2} \frac{\partial^2 c_{1,1} c_{2,1} U \left(\tilde{\Theta}_1^f \right)}{\left(\partial \tilde{I}_1 \right)^2}
\end{aligned}
\tag{115}
$$

where the $\Omega_n = \frac{\partial H_n}{\partial I_n}$ are the angular velocities of the unperturbed oscillators. For weak coupling, the k-dependent terms of Eq.(115) can be neglected, i.e.

$$
M \approx M_1 + M_2 := \Omega_1 \frac{\partial \Omega_1}{\partial H_1} + \Omega_2 \frac{\partial \Omega_2}{\partial H_2}
\tag{116}
$$

M_n quantifies the amplitude-frequency coupling of the unperturbed oscillators [45], which is zero for harmonic oscillators and in general is nonzero for nonlinear oscillators.

Higher order derivatives of the term $\bar{H}$ have been neglected in Eq.(114). Therefore, the variation of M^f has to be small, i.e. $\left| \frac{\partial M}{\partial \tilde{I}_1} \right|_{\tilde{I}_1^f} \Delta I \right| \ll \left| M^f \right|$. Due to Eq.(116), this condition is fulfilled if the variation of the amplitude-frequency coupling of both oscillators is small and M^f is large. Therefore at least one of the oscillators has to be nonlinear with large amplitude-frequency coupling. Furthermore the variation of the Fourier components has to be small in order to neglect the terms of order $k\Delta I$, i.e. $k \left| \frac{\partial c_{1,1} c_{2,1} U(\tilde{\Theta}_1^f)}{\partial \tilde{I}_1} \right|_{\tilde{I}_1^f} \right| \ll \left| M^f \Delta I \right|$.

The difference between the maximum and the minimum value of ΔI depends on the initial phase difference and is defined by

$$
\Delta I_d \left(\tilde{\Theta}_1^0 \right) := \max_t \left(\Delta I(t) \right) - \min_t \left(\Delta I(t) \right)
\tag{117}
$$

The dynamics defined by Eq.(114) is closely related to the dynamics of a pendulum (Fig.23), where $U^f \left(\tilde{\Theta}_1 \right)$ is a potential. For a large variety of nonlinear oscillators the

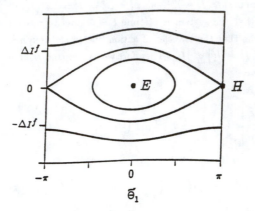

Figure 23: *The trajectories of two coupled oscillators in terms of the transferred action ΔI and their phase difference $\tilde{\Theta}$. The exchange of action is as large as possible for states in the neighborhood of the hyperbolic fixed point(H). The exchange of action is as small as possible at the elliptic fix point(E). The transfer of action is proportional to the transfer of energy.*

Fourier components $c_{n,l}$ fall off rapidly as l increases and $U^f\left(\tilde{\Theta}_1\right)$ has its absolute extrema at $\tilde{\Theta}_1^f = 0, \pm\pi$, i.e. at the stationary points. Then the maximum value ΔI_d^m of ΔI_d reads

$$
\begin{aligned}
\Delta I_d^m :=\ & \max_{\tilde{\Theta}_1^0}\left(\Delta I_d\left(\tilde{\Theta}_1^0\right)\right) = \\
=\ & \begin{cases} 2\Delta I^f & \text{for } |\Delta I^0| \leq \Delta I^f \quad \text{(separatrix)} \\ |\Delta I^0| - \left((\Delta I^0)^2 - \left(\Delta I^f\right)^2\right)^{\frac{1}{2}} & \text{else} \qquad\qquad \text{(rotation)} \end{cases}
\end{aligned} \tag{118}
$$

where ΔI^0 is the difference between the action at the fixed point and the action at $t = 0$, i.e. $\Delta I^0 := \tilde{I}_1^0 - \tilde{I}_1^f$. The magnitude of ΔI^0 can be estimated by

$$
|\Delta I^0| \approx \left|\frac{\omega_1^0 - \omega_2^0}{M^f}\right| \tag{119}
$$

and where ΔI^f is given by

$$
\Delta I^f = \left|\frac{kc_{1,1}^f c_{2,1}^f\left(U^f(0) - U^f(\pi)\right)}{M^f}\right|^{\frac{1}{2}} \tag{120}
$$

If one of the oscillators is harmonic or if the Fourier amplitudes $c_{n,l}$ fall off rapidly as l increases $\left|U^f(\pi) - U^f(0)\right| \approx 2$.

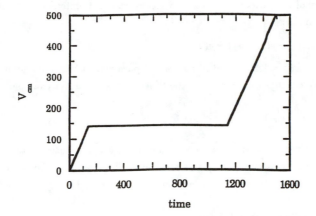

Figure 24: *The velocity of the center of mass seems to be attracted to the state where energy exchange between the oscillation of the center of mass motion and the vibration of the two particles x reaches a maximum.*

The maximum difference of the energy of each oscillator

$$\Delta H := \max_t (H_n(t)) - \min_t (H_n(t)) \tag{121}$$

depends on the initial condition $\tilde{\Theta}_1^0$. The maximum energy exchange $\Delta H^m := \max_{\tilde{\Theta}_1^0} \left(\Delta H \left(\tilde{\Theta}_1^0 \right) \right)$ can be estimated by

$$\Delta H^m \approx \omega^f \Delta I_d^m \tag{122}$$

whereas the energy exchange is zero at the minimum. It reaches those extrema at the states where $U^f \left(\tilde{\Theta}_1 \right)$ has its absolute extrema, i.e. at the stationary states.

L. ARSENAULT and others[82] are currently working on a generalization of this theorem for dissipative systems and more complicated nonlinear oscillators, since numerical simulations indicate that it is applicable in those systems, too. For example Fig.24 shows a conservative system consisting of two harmonically coupled particles in an inclined washboard potential. The corresponding equation of motion reads:

$$\ddot{x} = -2kx + 32\sin x_{cm} \cos x \tag{123}$$

$$\ddot{x}_{cm} = 1 + 32\cos x_{cm} \sin x \tag{124}$$

where x linear proportional to the distance between the two particles, x_{cm} is the location of the center of mass of the two particles and $k = 10^4$. Numerical simulation shows that the velocity of the center of mass remains approximately constant for a long period of time when the period of the up and down motion due to the wash board potential matches the eigenfrequency of the harmonic coupling. A generalization to interacting chaotic systems might read like this:

$$\dot{x}_i = f(x_i, p_i) + g(x_i - x_j) - g(0) \tag{125}$$

where $i = 1, 2, \ldots, j = 1, 2, \ldots$ and p is a control parameter which the system can adjust on a time scale which is slow compared the time scale of x_i, as, for example,

$$p_i(t) = \int_{t_T}^{t} h(x_i(t)) \tag{126}$$

f, g and h are given functions. $x_i(t) = x_j(t)$ for all i, j is a stationary state of the system where the interaction is extremal, in this case zero. This state corresponds to the elliptic fixed point of Fig.23. It is unknown whether the corresponding hyperbolic fixed point exists.

Another example are CHLADNY's dust figures on vibrating plates[83]. The dust collects mainly at the nodes of the vibration and a little bit at the maxima of the vibration, i.e. at those locations where the interaction between the vibrating plate and the motion of the dust particles is extremal.

5.3 The Principle of Matching, Modules, Adaptation, and the Evolution of
Control Parameters

The principle of the dynamical key states that only very special external driving forces provoke a large response in a nonlinear oscillator, which seems to make it very unlikely that a given natural system could stimulate a large response in another natural system or get any strong interaction with the other natural system. However, the principle of optimal interaction asserts that just those states with a complicated dynamics but an optimal interaction are stationary or even attractive. Therefore, we conclude that interacting physical systems are very likely to be observed or adapt to complicated states, where their interaction, for example, the size of the energy exchange, is as large as possible or as small as possible (principle of matching). The term "principle of matching" originates from Biology, where is has a very similar meaning[84].

According to the definition given in section 3.1, control parameters are those variables which have a time scale much longer than the typical time scale of the order parameters. The energy content of the two oscillators in section 5.2 is changing on a slow rate compared to the period of oscillation of each oscillator. Therefore, the energy content of each oscillator may be considered a slowly evolving control parameter. In this terminology the principle of matching reads: *Interacting physical systems adjust those control parameters which are not fixed by external constraints, in such a way that their interaction is extremal.*

The principle of matching would suggest that modules emerge if more than two physical systems interact, where a module is a set of strongly interacting systems and where

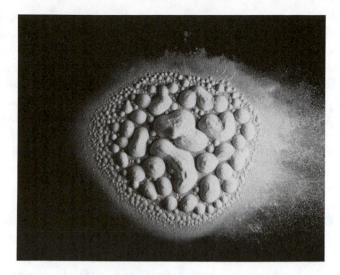

Figure 25: *Dust on a horizontal plate, which oscillates homogeneously up and down forms modules, i.e. sand piles separated by regions with nearly no dust. Dust on top of the sand piles are jumping only very short distances since the sand pile damps the vibration, whereas other particles travel at a high velocity.*

the boundary of a module is marked by weak interaction. Though little is known about these systems, some experiments seem to support this hypothesis(see Fig.25).

6. A VIRTUAL ENVIRONMENT FOR MATTER: SCIENCE FICTION?

In the context of the principle of matching we discussed adaptation of two physical systems. Now we want to pose the question about what happens if one of these physical systems is replaced by a computer. Digital computers are ideal chaos generators, since they are deterministic and react sensitively to any change in the computer code. As we have shown earlier, those chaotic signals can be used to provoke strong responses from physical systems. Sensors make it possible for computers to interact directly with matter and other physical systems. With these sensors the computer could establish an adapting virtual environment for physical systems. We can only guess in which way such an adapting virtual environment surrounding an adapting physical system would evolve. Maybe it would make it possible to produce completely relaxed, stress free, super clean materials or with a special type of disorder. Three things would be new: (i) the material is exposed to very different types of external nonequilibria as for example, acoustic and optic perturbations as well as a flow of heat and an electric current simultaneously, (ii) the external nonequilibria are not stationary but adapting, according to the response of the material and (iii) the external nonequilibria may have a complicated space and time dependence.

Even if such a virtual environment for matter is not visible in the near future, we expect that the paradigms which we presented in this paper might trigger a major change of scientific instrumentation and materials in the near future. Due to the principle of the dynamical key, function generators for aperiodic driving forces seem to be most useful. Since most function generators are microprocessor based, only minor changes are needed to upgrade a sine generator to a chaos generator. However, the replacement of this equipment in all scientific and industrial labs, for all frequency ranges, could be a major business.

Most nonlinear systems have many significant parameters. In order to determine those parameters through general resonance spectroscopy, it is necessary to scan through a high dimensional parameter space in search for the resonance, i.e. the measurement instrument needs to behave as a generic physical system, namely, adapt to the state of strongest interaction (principle of matching). It should be no technological problem to add an analog input to a standard function generator in order to use the feedback from the experimental system for a fast readjustment of the parameter setting of the function generator. This would upgrade a function generator to an adaptive system. Those fast adaptive function generators would make it possible to do an automatic search in high dimensional parameter spaces for a modeling of nonlinear systems.

Disordered and amorphous materials have many applications, for example, as high T_c superconductors. The spectroscopic method based on chaotic plane waves(sect. 4.2) makes it possible to check the quality of those materials extremely quickly and accurately. This may trigger some industrial innovations. The main consequence of the principle of matching is that systems in a strong external nonequilibrium tend to repair themselves if they are designed in such a way that the state where the system is supposed to operate is the state of largest interaction or smallest resistivity. Why are self repairing materials hardly being used today? Since they operate in strong external nonequilibrium, they tend to be complicated as a result of the nonequilibrium complexity paradigm. Consider, for example, the dendritic structures in section 5.1. Up to now there were no methods to characterize these complicated materials accurately, and this inaccuracy corrupted their reliability. This has changed now. The principle of the dynamical key offers many methods to determine the properties of complicated regenerative materials accurately. Since we expect products based on regenerative materials or systems, to have a substantially extended life time, regenerative materials could be a source for innovation in the future.

References

1. H. Haken, *Synergetics, an introduction* (Springer, Berlin, 1983), chapt. 7.

2. R.P. Lippmann, IEEE ASSP Mag., p. 4, April (1987).

3. L.A. Zadeh, *Fuzzy sets and applications* (Wiley, New York, 1987).

4. J.H. Holland, *Adaptation in natural and neural systems* (Univ. of Michigan Press, Ann Arbor, 1975).

5. C.G. Langton in *Artificial life*, edited by C.G. Langton. (Addison-Wesley, Redwood City, CA 1989).

6. H.G. Schuster, *Deterministic Chaos* (Physik-Verlag, Weinheim, 1984).

7. B.B. Mandelbrot, *The fractal geometry of nature* (Freeman, San Francisco, 1983).

8. D.K. Campbell, A.C. Newell, R.J. Schrieffer, H. Segur, Physica **18D**, 1(1986); R.K. Bullough and J.P. Caudrey eds., *Solitons* (Springer, New York, 1980).

9. R. Thom, *Structural stability and morphogenesis* (W.A. Benjamin, Reading, Mass. 1975).

10. P.C. Bak, C. Tang, and K. Wiesenfeld, Phys. Rev. A **38**, 364 (1988).

11. A. Hübler, E. Lüscher, Naturwissenschaften **76**, 67 (1989); A. Hübler, Helv. Phys. Acta **62**, 543 (1989).

12. K. Chang, A. Hübler, N. Packard in *Quantitative measures of complex dynamical systems*, edited by N.B. Abraham. (Plenum Press, New York, 1989), p. 225; K. Chang, S. Kodogeoriou, A. Hübler, E.A. Jackson, Physica D **44**, 407 (1990).

13. E.A. Jackson, A. Hübler, Physica D **44**, 407 (1990).

14. R. Shermer, A. Hübler, N. Packard, Phys. Rev. A **43**, 5642 (1991).

15. E. Ott, C. Grebogi, J.A. Yorke, Phys. Rev. Lett. **64**, 1196 (1990).

16. W.L. Ditto, S.N. Rauseo, M.L. Spano, Phys. Rev. Lett. **65**, 3211 (1990).

17. B. Plapp, A. Hübler, Phys. Rev. Lett. **65**, 2302 (1990).

18. T. Meyer, A. Hübler, N. Packard, *Quantitative measures of complex dynamical systems*, edited by N.B. Abraham. (Plenum Press, New York, 1989), p. 253.

19. G. Mayer-Kress, A. Hübler, in *Quantitative measures of complex dynamical systems*, edited by N.B. Abraham. (Plenum Press, New York, 1989), p. 155.

20. W.A. Brock in *Beyond belief: randomness, prediction, and explanation in science*, edited by J. Casti and A. Karlqvist. (CRC Press, Boca Raton, FL, 1991), p. 230.

21. M. Dueweke, A. Hübler, preprint 1991.

22. S. Aubry in *Seminar on the Rieman problem*, Lecture Notes in Mathematics (Springer, New York, 1980); E. Allroth, H. Müller-Krumbhaar, preprint 1983.

23. H. Haken, *Advanced Synergetics* (Springer, New York, 1983).

24. V.I. Arnold, *Mathematical methods of classical mechanics*, 4th ed. (Springer, New York, 1984), p. 4.

25. W. Stelzel, T. Kautzky, A. Hübler, E. Lüscher, Helv. Phys. Acta **61**, 224 (1988); R. Shermer, F. Dinkelacker, A. Hübler, V. Melnikov, preprint 91.

26. H. Haken, J.A.S. Kelso, H. Bunz, Biol. Cybern. **51**, 347 (1985); J.A.S. Kelso, G. Schöner, J.P. Scholz, H. Haken, Phys. Scripta **37**, 79 (1987).

27. T. Eisenhammer, A. Hübler, N. Packard, J.S.C. Kelso, Biological Cybernetics **65**, 107 (1991).

28. J.P. Crutchfield, K. Young, Phys. Rev. Lett. **63**, 105 (1989).

29. M.J. Feigenbaum, J. Stat. Phys. **19**, 25 (1978); J. Stat. Phys. **21**, 669 (1979).

30. J.D. Farmer, S. Sidorowich, Phys. Rev. Lett. **59**, 845 (1987).

31. J.P. Crutchfield, B.S. McNamara, Complex Syst. **1**, 417 (1987).

32. J. Cremers, A. Hübler, Z. Naturforsch. **42**a, 797 (1987).

33. A. Hübler, Helv. Phys. Acta **62**, 343 (1989); F. Ohle, F. Dinkelacker, A. Hübler, Technical Report CCSR-90-13, Center for Complex Systems Research, Univ. of Illinois at Urbana-Champaign.

34. A. Hübler, J. Miller, D. Pines, N. Weber, preprint 1991.

35. E. Roesch, H. Eckelmann, A. Hübler, Construction of Differential Equations from Experimental Data of Kármán Vortex Streets, report #11/1988, Max-Planck-Institut für Strömungsforschung, Göttingen, ISSN 0436-1199; F. Ohle, P. Lehmann, H. Eckelmann, A. Hübler, Phys. Fluids **A 2**, 479 (1990).

36. N.H. Packard, J.P. Crutchfield, J.D. Farmer, and R.S. Shaw, Phys. Rev. Lett. **45**, 712 (1980).

37. H. Whitney, Ann.Math. **37**, 645 (1936); **45**, 220 (1936), **45** 247 (1936).

38. F. Takens, in *Dynamical systems and turbulence*, edited by D.A. Rand and S. Young. (Springer, New York, 1981), p. 366.

39. T.P. Meyer, F.C. Richards, N. Packard, Phys. Rev. Lett. **63**, 1753 (1989).

40. J. Breeden, A. Hübler, Phys. Rev. A **42**, 5817 (1990); J. Breeden, F. Dinkelacker, A. Hübler, Phys. Rev. A **42**, 5827 (1990).

41. W. Eberl, T. Aschenbrenner, M.Athelogou, A. Hübler, preprint 1991.

42. G. Reiser, A. Hübler, E. Lüscher, Z. Naturforsch. **42a**, 803 (1987).

43. B.A.Huberman and J.P.Crutchfield, Phys. Rev. Lett. **43**, 1743 (1979); D.D. Humieres, M.R. Beasley, B.A. Huberman, and A. Libchaber, Phys. Rev. A **26**, 2483 (1982).

44. D.Ruelle, Phys. Rev. Lett. **56**, 405 (1986); U.Parlitz and W. Lauterborn, Phys. Lett. **107A**, 351 (1986).

45. A.H. Nayfeh and D.T. Mook, *Nonlinear oscillations* (John Wiley & Sons, New York, 1976), chapt. 4.1.

46. T. Eisenhammer, A. Hübler, T. Geisel, E. Lüscher, Phys. Rev. A **41**, 3332 (1990).

47. A.J. Lichtenberg and M.A. Lieberman, *Regular and stochastic motion* (Springer, New York, 1982), chapt.3/4.

48. D. Bensen, M. Welge, A. Hübler, N. Packard, in *Quantitative Measures of Complex Dynamical Systems*, edited by N.B. Abraham. (Plenum Press, New York, 1989), p. 209.

49. K. Chang, A. Hübler, preprint 1991.

50. S. A. Kauffman, private communication.

51. J. Wang, A. Hübler, preprint 1991.

52. S. Krempl, A. Hübler, preprint 1991.

53. R. Wittmann, M. Mondello, G. Bertsos, A. Hübler, preprint(1991).

54. S. Shi, H. Rabitz, J. Chem. Phys **92**, 2972 (1990); S. Shi, H. Rabitz, J. Chem. Phys **92**, 364 (1990).

55. S. Chelkowski, A.D. Bandrauk, Phys. Rev. Lett. **65**, 2355 (1990); S. Chelkowski, A.D. Bandrauk, Phys. Rev. A **41**, 6480 (1990).

56. T. Eisenhammer, A. Hübler, G. Mayer-Kress, P.W. Milonni, to be published.

57. M.E. Goggin, P.W. Milonni, Phys. Rev. A **37**, 796 (1987).

58. C. Rolland, P.B. Corcum, J. Opt. Soc. Am. **5B**, 641 (1988).

59. W.H. Knox, R.L. Fork, M.C. Downer, R.H. Stolen, C.U. Shank, Appl. Phys. Lett. **46**, 1120 (1985).

60. R.B. Walker, R.K. Preston, J. Chem. Phys. **67**, 2017 (1977).

61. P.B. Corcum, N.H. Burnett, F. Brunel, Phys. Rev. Lett. **62**, 1259 (1989).

62. S. Krempl, A. Hübler, to be published.

63. D. Ruelle, Physica **A113**, 619 (1982).

64. P. Bak, Rep. Prog. Phys. **45**, 587 (1982).

65. P. Reichert, R. Schilling, Phys. Rev. **B30**, 917 (1984).

66. S. Martin, W. Martienssen, Z. Phys. **B68**, 299 (1987).

67. G. Fritsch, Naturwissenschaften **75**, 551 (1988); E. Lüscher, G. Fritsch, G. Jacucci (Eds.), Amorphous and Liquid Metals (NATO ASI, E118, Martinus Nijhoff, Dordrecht, Boston, Lancaster, 1987).

68. P. Reichert, R. Schilling, Phys. Rev. **B32**, 5731 (1985).

69. J.-P. Sadoc, J. de Phys. (Paris) **41**, Coll. C8, 326 (1980); J.-P. Sadoc, R. Mosseri, Phil. Mag. **B45**, 467 (1982).

70. P.M. Chaikin, A. Behrooz, M.A. Itzler, C. Wilks, B. Whitehead, G. Grest, D. Levine, Physica **B152**, 113 (1988).

71. G. Mayer-Kress, Zur Persistenz von Chaos und Ordnung, Ph.D. thesis, Institut für Theoretische Physik und Synergetik, Universität Stuttgart 1984, S. Beckert, U. Schock, C.D. Schultz, T. Weidlich, F. Kaiser, Phys. Lett. 107A, 304 (1985).

72. A. Hübler, R. Georgii, M. Kuchler, W. Stelzl, E. Lüscher, Helv. Phys. Acta **61**, 897 (1988); R. Georgii, W. Eberl, E. Lüscher, A. Hübler, Helv. Phys. Acta **62**, 290 (1989).

73. M. Rose, T. Kautzky, P. Deisz, A. Hübler, E. Lüscher, Helv. Phys. Acta **62**, 286 (1989).

74. see for example B.D.O. Anderson, *Stability Analysis of Adaptive Systems: Passivity and Averaging Analysis*, Cambridge, MA, MIT Press 1986.

75. A.Hübler, Ph.D. thesis at Technische Universität München (1987).

76. P. Glansdorff, I. Prigogine, *Thermodynamic theory of structure, stability and fluctuations* (Wiley, New York, 1971).

77. B. Merté , G. Hadwich, B. Binias, P. Deisz, A. Hübler, E. Lüscher, Helv. Phys. Acta **62**, 294 (1989); M. Athelogou, P. Deisz, B. Merté , A. Hübler, E. Lüscher, Helv. Phys. Acta **62**, 250 (1989).

78. M.E. Goggin, P.W. Milonni, Phys. Rev. **A 37** 796 (1988); G.R. Smith, A.N. Kaufmann, Phys. Rev. Lett. **34** 1613 (1975); B.V. Chirikov, Proc. R. Soc. London **A 413** 145 (1987); R.V. Jensen, Phys.Rev. **A30** 386 (1984); J.G. Leopold, D. Richards, J.Phys. **B19** 1125 (1986).

79. A.J. Lichtenberg, M.A. Lieberman, op.ct., p.100ff

80. N.N. Bogoliubov, Y.A. Mitropolsky, *Asymptotic methods in the theory of nonlinear oscillations* (Gordon Breach, New York).

81. M.D. Kruskal, J. Math. Phys **3** 806 (1962)

82. L. Arsenault, A. Hübler, preprint 1991.

83. F. Dinkelacker, A. Hübler, and E. Lüscher, Biological Cybernetics **56**, 51 (1987).

84. J. E. Mittenthal, A. B. Baskin, in *Principles of organization in organisms* edited by J.E. Mittenthal and A. B. Baskin (Addison-Wesley, Redwood City, CA, 1992).

Knowledge and Meaning: Chaos and Complexity

J.P. Crutchfield

What are models good for? Taking a largely pedagogical view, the following essay discusses the semantics of measurement and the uses to which an observer can put knowledge of a process's structure. To do this in as concrete a way as possible, it first reviews the reconstruction of probabilistic finite automata from time series of stochastic and chaotic processes. It investigates the convergence of an observer's knowledge of the process's state; assuming that the process is in, and the observer also uses for internal representation, that model class. The conventional notions of phase and phase-locking are extended beyond periodic behavior to include deterministic chaotic processes. The meaning of individual measurements of an unpredictable process is then defined in terms of the computational structure in a model that an observer built.

1. Experimental Epistemology

The grammar of a language can be viewed as a theory of the structure of this language. Any scientific theory is based on a certain finite set of observations and, by establishing general laws stated in terms of certain hypothetical constructs, it attempts to account for these observations, to show how they are interrelated and to predict an indefinite number of new phenomena.

General linguistic theory can be viewed as a metatheory which is concerned with the problem of how to choose such a grammar in the case of each particular language on the basis of a finite corpus of sentences.

N. Chomsky[1]

There is a deep internal conflict in the scientific description of a classical universe. The first aspect of this conflict is the conviction that nature is, in a sense, a deductive system: pushing forward through time according to the dictates of microscopic deterministic equations of

---

Physics Department, University of California, Berkeley, California 94720, USA; For Internet use chaos@gojira.berkeley.edu or try chaos@UCBphysi.BITNET.

motion. The second, polar aspect is that a scientist, or any physical subsystem of the universe which attempts to model other portions of the universe, is an inferential system: destined always to estimate and approximate from finite data and, for better or worse, project unverifiable structure onto the local environment. The conflict is simply that the logical types of these concomitant views do not match. Presumably due to the difficulty represented by this tension, the premises underlying scientific description leave unanswered the primary questions: Of what does knowledge consist? And, from what does meaning emerge?

Answers to questions like these usually fall under the purview of the philosophy of science.[2] Unfortunately, investigations along these lines tend to be more concerned with developing an internally-consistent language for the activity of science than with testing concrete hypotheses about the structure of knowledge and the mechanisms governing the emergence of meaning in physical systems. Retreating some distance from the general setting of epistemology and ontology, the following attempts to address these questions in the philosophically restricted domain of nonlinear modeling. The presentation is narrowed, in fact, even further to building stochastic automata from time series of finite resolution measurements of nonlinear stationary processes. Nonetheless, I think that some progress is made by introducing these issues into modeling complex phenomena and that the results reflect back positively on the larger problems.

Why are semantics and epistemology of interest to the study of chaos and complexity? Surely such philosophical questions have arisen before for mechanical systems. The early days of information theory, cybernetics, control theory, and linguistics, come to mind. Perhaps the most direct early approach is found in MacKay's collection **Information, Meaning and Mechanism**.[3] But I find the questions especially compelling, and even a bit poignant, in light of the recent developments in dynamics. We now know of easily-expressed nonlinear systems to whose temporal behavior we ascribe substantial complication. In stark opposition to Laplacian determinism, even if we know their governing equations of motion, they can remain irreducibly unpredictable. The behavior's complication and its unpredictability contrasts sharply with the simplicity of the equations of motion.* And so, in the microcosm of nonlinear modeling a similar epistemological tension arises: the deterministic evolution of simple systems gives rise to complex phenomena in which an observer is to find regularity.

I should also mention another premise guiding these concerns. A goal here is to aid investigations in the dynamics of evolution. The premise is that the questions of experimental epistemology and the emergence of semantics must be addressed at the very low level of the following. Additionally, these issues must be tackled head on if we are to understand

---

* This is also testimony to how much we still do not understand about how genuine complexity is generated by dynamical systems such as the oft-studied logistic and circle maps. The engineering suggestions[4–7] that there exist physically plausible dynamical systems implementing Turing machines, though they appear to address the problem of physical complexity, are irrelevant as they skirt the issue of its emergence through time evolution.

genuine learning and intrinsic self-organization beyond mere statistical parameter estimation and pattern stabilization, with which they respectively are often confused.

The epistemological problem of nonlinear modeling is: Have we discovered something in our data or have we projected the new-found structure onto it?* This was the main lesson of attempting to reconstruct equations of motion from a time series:[9] When it works, it works; when it doesn't, you don't know what to do; and in both cases it is ambiguous what you have learned. Even though data was generated by well-behaved, smooth dynamical systems, there was an extreme sensitivity to the assumed model class that completely swamped "model order estimation". Worse still there was no *a priori* way to select the class appropriate to the process.† This should be contrasted with what is probably one of the more important practical results in statistical modeling: within a model class a procedure exists to find, given a finite amount of data, an optimal model that balances prediction error against model complexity.[10] Despite representations to the contrary, this "model order estimation" procedure does not address issues of class inappropriateness and what to do when confronted with failure.

There appears to be a way out of the model class discovery dilemma. The answer that hierarchical machine reconstruction gives is to start at the lowest level of representation, the given discrete data, and to build adaptively a series of models within a series of model classes of increasing computational capability until a finite causal model is found.[11] Within each level there is a model-order-estimation inference of optimal models, as just indicated. And there is an induction from a series of approximate models within a lower "inappropriate" class to the next higher model class. This additional inductive step, beyond the standard model order estimation procedure, is the price to be paid for formalizing adaptive modeling.

The success at each stage in hierarchical reconstruction is controlled by the amount of given data, since this puts an upper bound on statistical accuracy, and an error threshold, which is largely determined by the observer's available computational resources. The goal is to find a finite causal model of minimal size and prediction error while maximizing the extraction of information from the given data.

Within this hierarchical view the epistemological problem of nonlinear modeling can be crudely summarized as the dichotomy between engineering and science. As long as a representation is effective for a task, an engineer does not care what it implies about the underlying mechanisms; to the scientist though the implication makes all the difference in the world. The engineer certainly is concerned with minimizing implementation cost, such as representation size, compute time, and storage; but the scientist presumes, at least, to be focused on what the model means *vis á vis* natural laws. The engineering view of science is that it is mere data compression; scientists seem to be motivated by more than this.

---

* In philosophy such considerations currently appear under the rubric of "intentionality".[8] I do not have in mind, however, verbal intercourse, but rather the intentionality arising in the dialogue, and sometimes monologue, between science and nature. One of the best examples of this is the first "law" of thermodynamics: energy conservation.

† In artificial intelligence this is referred to as the "representation problem".

From these general considerations a number of pathways lead to interesting and basic problems. The following will address the questions of what individual measurements mean to the observer who has made them and how an observer comes to know the state of a process. There are two main themes: knowledge convergence and measurement semantics. While these are relatively simple compared to the problems of an experimental epistemology, they do shed some light on where to begin and so complement direct studies of the emergence of intrinsic computation. Before the two themes are considered I quickly review stochastic automata and their reconstruction. Indeed, to complete the semantic chain of meaning and to suggest how it can emerge, I must say first where models come from. Then, after the main topics are presented, the discussion ends with some general remarks on causality and complexity.

2. COMPUTATIONAL MECHANICS

The overall goal is to infer from a series of measurements of a process a model of the generating mechanism. Additionally, the model is to indicate the process's computational structure. This refers not only to its statistical properties, such as the decay of correlation, but also to the amount of memory it contains and, for example, whether or not it is capable of producing the digit string of (say) $\sqrt{3}$. The extraction of these properties from the model determines the utility of the model class beyond mere temporal prediction of the process's behavior.

The first subsection starts off by defining the character of the data from which the model is built. The next two subsections then review machine reconstruction and the statistical mechanics of the inferred stochastic machines. The section closes with some comments on complexity and model minimality.

2.1 The Measurement Channel

The universe of discourse for nonlinear modeling consists of a process P, the measuring apparatus $\mathcal{I}$, and the modeler itself. Their relationships and components are shown schematically in Fig. 1. The goal is for the modeler, by taking advantage of its available resources, to make the "best" representation of the nonlinear process.

The process P, the object of the modeler's ultimate attention, is the unknown, but hopefully knowable, variable in this picture. And so there is little to say, except that it can be viewed as governed by stochastic evolution equations

$$\vec{X}_{t+\Delta t} = \vec{F}\left(\vec{X}_t, \vec{\xi}_t, t\right) \tag{1}$$

where $\vec{X}_t$ is the configuration at time t, $\vec{\xi}_t$ some noise process, and $\vec{F}$ the governing equations of motion. The following discussion also will have occasion to refer to the process's measure $\mu\left(\vec{X}\right)$ on its configuration space and the entropy rate $h_\mu\left(\vec{X}\right)$ at which it produces information.

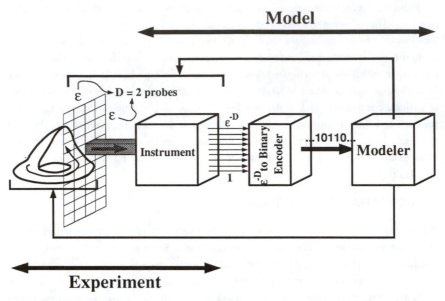

Fig. 1 The Big Channel. The flow of information (measurements) on the shortest time scales is from the left, from the underlying process, to the right toward the modeler. The latter's task is to build the "best" representation given the available data set and computational resources. On longer time scales the modeler may modify the measuring apparatus and vary the experimental controls on the process. These actions are represented by the left-going arrows. Notice that from the modeler's perspective there is a region of ambiguity between the model and the experiment. The model includes the measuring apparatus since it instantiates many of the modeler's biases toward what is worth observing. But the experiment also includes the measuring apparatus since it couples to the process. Additionally, the apparatus is itself a physical device with its own internal dynamics of which the modeler may be unaware or incapable of controlling.

The measuring apparatus is a transducer that maps $\vec{X}_t$ to some accessible states of an instrument $\mathcal{I}$. This instrument has a number of characteristics, most of which should be under the modeler's control. The primary interaction between the instrument and the process is through the measurement space $\mathcal{R}^D$ which is a projection $\mathcal{P}$ of $\vec{X}_t$ onto (say) a Euclidean space whose dimension is given by the number D of experimental probes. An instrument that distinguishes the projected states to within a resolution ϵ partitions the measurement space into a set $\Pi_\epsilon(D) = \left\{ \pi_i : \pi_i \subset \mathcal{R}^D, i = 0, \dots \epsilon^{-D} \right\}$ of cells. Each cell π_i is the equivalence class of projected states that are indistinguishable using that instrument. The instrument represents the event of finding $\mathcal{P}\left(\vec{X}_t\right) \in \pi_i$ by the cell's label i. With neither loss of generality nor information, these indices are then encoded into a time-serial binary code. As each measurement is made its code is output into the data stream. In this way, a time series of measurements made by the instrument becomes a binary string, the data stream s available to the modeler. This is a discretized set of symbols $\mathbf{s} = \dots s_{-4}s_{-3}s_{-2}s_{-1}s_0s_1s_2s_3s_4 \dots$ where in a single measurement made by the modeler the instrument returns a symbol $s_t \in \mathbf{A}$ in an alphabet $\mathbf{A}$ at time index $t \in \mathbf{Z}$. Here we take a binary alphabet $\mathbf{A} = \{0, 1\}$.

Beyond the instrument, one must consider what can and should be done with information in the data stream. Acquisition of, processing, and inferring from the measurement sequence are the functions of the modeler. The modeler is essentially defined in terms of its available inference resources. These are dominated by storage capacity and computational power, but certainly include the inference method's efficacy, for example. Delineating these resources constitutes the barest outline of an observer that builds models. Although the following discussion does not require further development at this abstract a level, it is useful to keep in mind since particular choices for these elements will be presented.

The modeler is presented with s, the bit string, some properties of which were just given. The modeler's concern is to go from it to a useful representation. To do this the modeler needs a notion of the process's effective state and its effective equations of motion. Having built a model consisting of these two components, any residual error or deviation from the behavior described by the model can be used to estimate the process's effective noise level. This level is determined by the amount of data "unexplained" by the inferred model. It should be clear when said this way that the noise level and the model's sophistication depend directly on the data and on the modeler's resources. Finally, the modeler may have access to experimental control parameters. And these can be used to aid in obtaining different data streams useful in improving the model by (say) concentrating on behavior where the effective noise level is highest.

The central problem of nonlinear modeling now can be stated. Given an instrument, some number of measurements, and fixed *finite* inference resources, how much computational structure in the underlying process can be extracted?

Before pursuing this goal directly it will be helpful to point out several limitations imposed by the data or, rather, the moder's interpretation of it.

In describing the data stream's character it was emphasized that the individual measurements are only indirect representations of the process's state. If the modeler interprets the measurements as the process's state, then it is unwittingly forced into a class of computationally less powerful representations. These consist of finite Markov chains with states in $\mathbf{A}$ or in some arbitrarily selected state alphabet. This will become clearer through several examples used later on. The point is that it is important to not over-interpret the measurements' content at this early stage of inference as this might limit the quality of the resulting models.

The instrument itself obviously constrains the observer's ability to extract regularity from the data stream and so it directly affects the model's utility. The most basic of these constraints are given by Shannon's coding theorems.[12] The instrument was described as a transducer, but it also can be considered to be a communication channel between the process and the modeler. The capacity of this channel in bits per measurement is $\dot{I} = \tau^{-1} H(\Pi_\epsilon(\mathrm{D}))$, where $H(\{p_i\}) = -\sum_{\{p_i\}} p_i \log_2 p_i$ is the Shannon entropy of the distribution $\{p_i = \mu(\pi_i) : \pi_i \in \Pi_\epsilon\}$. If the process is deterministic and has entropy $h_\mu\left(\vec{X}\right) > 0$, a theorem of Kolmogorov's says that this rate is maximized for a given process if the partition

$\Pi_\epsilon(D)$ is generating.[13] This property requires infinite sequences of cell indices to be in a finite-to-one correspondence with the process's states. A similar result was shown to hold for the classes of process of interest here: deterministic, but coupled to an extrinsic noise source.[14] Note that the generating partition requirement necessarily determines the number D of probes required by the instrument.

This is, however, the mathematical side of the problem here: verification of a generating partition requires knowledge of the equations of motion. Thus, Kolmogorov's result turns the inference problem on its head. Here interest is focused on the opposite, pragmatic problem of discovering structure from given data. As indicated in [14], the notion of generating partitions gives an operational definition of a good instrument that is useful if the instrument can be altered. If the data is simply given, one can still proceed with modeling. The conclusions cannot be as strong, though, as when the equations of motion are known. Then, again, such knowledge is an idealization.

For an instrument with a generating partition, Shannon's noiseless coding theorem says that the measurement channel must have a capacity higher than process's entropy

$$\dot{I} \geq h_\mu\left(\vec{X}\right) \tag{2}$$

If this is the case then the modeler can use the data stream to reconstruct a model of the process and, for example, estimate its entropy and complexity. These can be obtained to within error levels determined by the process's extrinsic noise level and the number of measurements.

If $\dot{I} < h_\mu\left(\vec{X}\right)$, then Shannon's theorems say that the modeler will not be able to reconstruct a model with an effective noise level less than the equivocation $h_\mu\left(\vec{X}\right) - \dot{I}$ induced by the instrument. That is, there will be an "unreconstructable" portion of the dynamics represented in the signal.

These results assume, as is also done implicitly in Shannon's existence proofs for codes, that the modeler has access to arbitrary inference resources. When these are limited there will be yet another corresponding loss in the quality of the model and an increase in the apparent noise level. It is interesting to note in this context that if one were to adopt Laplace's philosophical stance that all (classical) reality is deterministic and update it with the modern view that it is chaotic, then the inferential limitations discussed here are the general case. Apparent randomness is a consequence of them and not a property of unobserved nature.

2.2 Computation from a Time Series

On what sort of structure in the data stream should the models be based? If the goal includes prediction, as the preceding assumed, then a natural object to reconstruct from the data series is a representation of the process's instantaneous state. Unfortunately, as already noted, individual measurements are only indirect representations of the state. Indeed, the instrument simply may not supply data of a quality sufficient to discover the true states, independent of the amount of data. So how can the process's "effective" states be accessed?

The answer to this turns on a generalization of the "reconstructed states" introduced, under the assumption that the process is a continuous-state dynamical system, by Packard *et al.*[15] The contention there was that a single time series necessarily contained all of the information about the dynamics of that time series. The notion of reconstructed state was based on Poincaré's view of the intrinsic dimension of an object.[16] This was defined as the largest number of successive cuts through the object resulting in isolated points. A spherical shell in three dimensions by his method is two dimensional since the first cut typically results in a circle and then a second cut, of that circle, isolates two points. One way Packard *et al.* implemented this used probability distributions conditioned on values of the time series' derivatives. That is, the coordinates of the reconstructed state space were taken to be successive time derivatives and the cuts were specified by setting their values. This was, in fact, an implementation of the differential geometric view of the derivatives as locally spanning the graph of the dynamic.

In this reconstruction procedure a state of the underlying process is identified by increasing the number of conditioning variables, i.e. employing successively higher derivatives, until the conditional probability distribution peaks. It was noted shortly thereafter that in the presence of extrinsic noise a number of conditions is reached beyond which the conditional distribution is no longer sharpened.[14] And, as a result the process's state cannot be further identified. The width of the resulting distribution then gives an estimate of the effective extrinsic noise level and so also an estimate of the maximum amount of information contained in observable states. The minimum number of conditions first leading to this situation is an estimate of the effective dimension.

The method of time derivative reconstruction gives the key to discovering states in discrete times series. For discrete time series a state is defined to be the set of subsequences that render the future conditionally independent of the past.[17]* Thus, the observer identifies a state at different times in the data stream as its being in identical conditions of ignorance about the future. (See Fig. 2.)

Let's introduce some notation here. Consider two parts of the data stream $\mathbf{s} = \ldots s_{-2}s_{-1}s_0s_1s_2\ldots$. The one-sided forward sequence $\mathbf{s}_t^{\rightarrow} = s_t s_{t+1} s_{t+2} s_{t+3} \ldots$ and one-sided reverse sequence $\mathbf{s}_t^{\leftarrow} = \ldots s_{t-3} s_{t-2} s_{t-1} s_t$ are obtained from $\mathbf{s}$ by splitting it at time t into the forward- and reverse-time semi-infinite subsequences. Consider the joint distribution of possible forward sequences $\{\mathbf{s}^{\rightarrow}\}$ and reverse sequences $\{\mathbf{s}^{\leftarrow}\}$ over all times t

$$Pr(\mathbf{s}^{\rightarrow}, \mathbf{s}^{\leftarrow}) = Pr(\mathbf{s}^{\rightarrow}|\mathbf{s}^{\leftarrow})Pr(\mathbf{s}^{\leftarrow}) \tag{3}$$

The conditional distribution $Pr(\mathbf{s}^{\rightarrow}|\omega)$ is to be understood as a function over all possible forward sequences $\{\mathbf{s}^{\rightarrow}\}$ that can follow the particular sequence ω where ever it occurs in $\mathbf{s}$.

---

* This notion of state is widespread; appearing in various guises in early symbolic dynamics and ergodic and automata theories. It is close to the basic notion of state in Markov chain theory. Interestingly, the notion of conditional independence is playing an increasingly central role in the design of expert systems.[18]

74

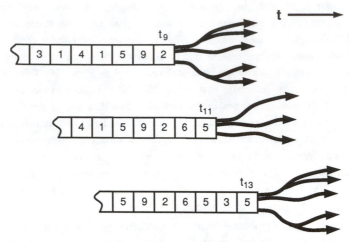

Fig. 2 Morph-equivalence induces conditionally independent states. When the template of future possibilities, i.e. allowed future subsequences and their past-conditioned probabilities, has the same structure then the process is in the same generalized state. At t_9 and at t_{13}, the process is in the same state; at t_{11} it is in another different state.

Then the same state $S \in \mathbf{S}$ is associated with all those times $t, t' \in \{t_{i_1}, t_{i_2}, t_{i_3} \ldots : i_k \in \mathbf{Z}\}$ such that past-conditioned future distributions are the same. That is,

$$t \sim t' \text{ if and only if } Pr(\mathbf{s}^{\rightarrow}|\mathbf{s}_t^{\leftarrow}) = Pr(\mathbf{s}^{\rightarrow}|\mathbf{s}_{t'}^{\leftarrow}) \tag{4}$$

If the source generating the data stream is ergodic, then there are several comments that serve to clarify how this relation defines states. First, the sequences $\mathbf{s}_t^{\leftarrow}$ and $\mathbf{s}_{t'}^{\leftarrow}$ are typically distinct. If $t \sim t'$, Eq. (4) means that upon having seen different histories one can be, nonetheless, in the same state of anticipation or ignorance about what will happen in the future. Second, $\mathbf{s}_t^{\leftarrow}$ and $\mathbf{s}_{t'}^{\leftarrow}$, when considered as particular symbol sequences, will each occur in $\mathbf{s}$ many times other than t and t', respectively. Finally, the conditional distributions $Pr(\mathbf{s}^{\rightarrow}|\mathbf{s}_t^{\leftarrow})$ and $Pr(\mathbf{s}^{\rightarrow}|\mathbf{s}_{t'}^{\leftarrow})$ are functions over a nontrivial range of "follower" sequences $\mathbf{s}^{\rightarrow}$.

This gives a formal definition to the set $\mathbf{S}$ of states as equivalence classes of future predictability: $\sim$ is the underlying equivalence relation that partitions temporal shifts of the data stream into equivalence classes. The states are simply labels for those classes. For a given state S the set of future sequences $\{\mathbf{s}_S^{\rightarrow} : S \in \mathbf{S}\}$ that can be observed from it is called its future morph. The set of sequences that lead to S is called its past morph. Anticipating a later section somewhat, note that the state and its morphs are the contexts in which an individual measurement takes on semantic content. Each measurement is anticipated or "understood" by the observer *vis á vis* its model and in particular the structure of the states. Once these states are found, the temporal evolution of the process, its (symbolic) dynamic, is given by a mapping from states to states $T : \mathbf{S} \rightarrow \mathbf{S}$; that is, $S_{t+1} = TS_t$.

The available reconstruction algorithms infer the states **S** via various approximations of the equivalence class conditions specified in (4).[11,17,19] I refer to these procedures generically as "machine reconstruction". The result of machine reconstruction, then, is the discovery of the underlying process's "hidden" states. This should be contrasted with the *ad hoc* methods employed in hidden Markov modeling in which a set of states and a transition structure are imposed by the modeler at the outset.[20,21]

Thus, the overall procedure has two steps. This first is to identify the states and the second is to infer the transformation T. In the following I review the simplest implementation since this affords the most direct means of commenting on certain properties of the resulting representations.

Initially, a parse tree is built from the data stream **s**. A window of width D is advanced through **s** one symbol at a time. If $s = \ldots 1010101110101 \ldots$ then at some time t the $D = 5$ subsequence $s_t^5 = 10101$ is seen, followed by $s_{t+1}^5 = 01010$. Each such subsequence is represented in the parse tree as a path. The tree has depth $D = 5$. (See Fig. 3).

Counts are accumulated at each tree node as each subsequence is put into to the parse tree. If the associated path is not present in the tree, it is added; new nodes each begin with a count of 1 and counts in existing nodes are incremented. If **s** has length N, then a node probability, which is also the probability of the sequence leading to it, is estimated by its relative frequency

$$Pr(n) = \frac{c_n}{N - D} \qquad (5)$$

where c_n is the count accumulated at node n. The node-to-node transition probabilities $Pr(n \to n')$ are estimated by

$$Pr(n \to n') = Pr\left(s|s^k\right) = \begin{cases} \frac{Pr(ss^k)}{Pr(s^k)} \approx \frac{c_{n'}}{c_n} & c_n > 0 \\ 0 & \text{otherwise} \end{cases} \qquad (6)$$

where the length k sequence s^k leads to node n and the length $k+1$ sequence $s^{k+1} = ss^k$ leads to node n' on symbol $s \in \mathbf{A}$.

The window length D is an approximation parameter. The longer it is, the more correlation, structure, and so on, is captured by the tree. Since **s** is of finite length the tree depth cannot be indefinitely extended. Thus, to obtain good estimates of the subsequences generated by the process and of the tree structure, the tree depth must be set at some optimal value. That value in turn depends on the amount of available data. The procedure for finding the optimal depth D is discussed elsewhere.

To infer the set of states we look for distinct subtrees of a given depth L. This is the length of future subsequences over which the morphs must agree. This step introduces a second approximation parameter, the morph depth L. Naturally, $L \leq D$ and typically one takes $2L = D$. Over a range of sources and for a given tree depth, that choice balances two requirements for the morph's optimal statistical estimation. The first is the need for a large number of examples of any given morph. This suggests taking $L \ll D$ since the

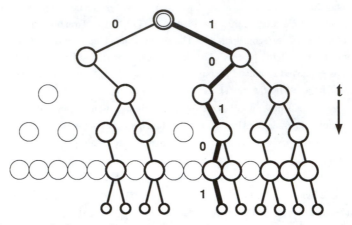

Fig. 3 The parse tree for the process "every other symbol is a 1". The tree is a hierarchical representation of the subsequences produced by the process. Here the tree depth is $D = 5$. If $\mathbf{s} = \ldots 1010101110101 \ldots$, then the sequence $s_t^5 = 10101$ will be seen. This is put into the parse tree, as shown by the bold line. Note that time goes down the tree. When counts of the subsequence that lead to a given tree node are accumulated, the tree gives a hierarchical representation of the infinite sequence probability distribution.

upper bound on the number of morph instances is $2^{D-L+1} - 1$. The second requirement is to allow a sufficient diversity of morphs; and so of states. This inclines one toward taking $L \approx D$, since the number of distinct depth L subtrees is $2^{2^L} - 1$. The analysis of this optimization is presented elsewhere. Investigating the parse tree of Fig. 3 using depth $L = 2$ subtrees one finds the three morphs shown in Fig. 4. In this way, three "machine" states $\mathbf{S} = \{A, B, C\}$ have been discovered for the "every other symbol is a 1" process of Fig. 3 up to the approximation implied by setting $D = 5$ and $L = 2$. It can be shown that these are sufficient to infer an exact model of the process.

The state-to-state transition structure is obtained by looking at how the morphs change into one another upon moving down the parse tree, i.e. upon reading $s = 0$ or $s = 1$. The resulting transformation T is represented graphically by the machine shown in Fig. 5. This should be compared to the parse tree (Fig. 3). There the morph below the top node, which is associated with machine state A, makes a transition on a 1 to a tree node with a morph below it in the same equivalence class (machine state A). On a 0, though, the top tree node makes a transition to a tree node with a morph associated with machine state B. In just this way the machine of Fig. 5 summarizes the morph to morph transition structure T on the parse tree.

This exposition covered only topological reconstruction: subtrees were compared only up to the subsequences $\{s_t^{\rightarrow}\}$ which were observed. Of course, what is needed are states that are not only topologically distinct, but also distinct in probability as indicated by (4). For this we must choose some metric to compare the conditional histograms associated with the morphs. I will briefly sketch one way to do this.

The goal is to develop an implementable approximation to (4) which defines states in terms of tree node conditional probability equivalence classes. Those equivalence classes

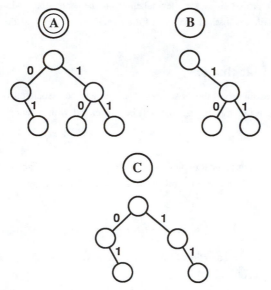

Fig. 4 The morph depth $L = 2$ subtrees found below tree nodes down to depth 3 in Fig. 3's parse tree. Each morph, or subtree, has been labeled by its associated machine state (cf. Fig. 5).

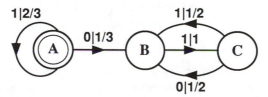

Fig. 5 The (topologically) reconstructed machine for the process "every other symbol is a 1".

require a subtree similarity relation between two arbitrary tree nodes; denote these n and n'. First assume that these nodes are topologically similar; that is, their morphs m_n and $m_{n'}$ are the same, up to some morph depth L,

$$n \underset{L}{\sim} n' \text{ if and only if } m_n = m_{n'}$$
$$\text{where } m_n = \left\{ s^L | n \right\} \text{ and } m_{n'} = \left\{ s^L | n' \right\} \tag{7}$$

where s^L is a length L sequence. If $m_n \neq m_{n'}$, then certainly the nodes cannot be similar in probability. Probabilistic similarity is then an additional constraint and defined by

$$n \underset{\delta}{\sim} n' \text{ if and only if } \left| Pr(\omega|n) - Pr(\omega|n') \right| \leq \delta, \ \forall \omega \in m_n \tag{8}$$

δ is yet another approximation parameter. Without going into details, it is important to note that the implied comparisons for subtree similarity can be done in a relatively efficient recursive algorithm that stops as soon as a difference in the morph structure or node transition probabilities is found at some level in the two subtrees under comparison. Finally, the state to state transition probabilities are then taken from the estimated node-to-node transition probabilities. Detailed statistical analysis of the overall procedure and the optimal selection of δ is given elsewhere.

2.3 Statistical Mechanics

Machine reconstruction gives then a set of states, that will be associated with a set $\mathbf{V} = \{v\}$ of vertices, and a set of transitions, that will be associated with a set

$$\mathbf{E} = \left\{ e : e \sim v \underset{s}{\rightarrow} v', \quad v, v' \in \mathbf{V}, s \in \mathbf{A} \right\} \tag{9}$$

of labeled edges. The graphical depiction of $M = \{\mathbf{V}, \mathbf{E}, \mathbf{A}\}$ is a labeled directed graph as seen above in Fig. 5. The full probabilistic structure is described by a set of input-alphabet labeled transition matrices

$$\left\{ T^{(s)} : \left(T^{(s)} \right)_{vv'} = p_{v \underset{s}{\rightarrow} v'}, \quad v, v' \in \mathbf{V}, s \in \mathbf{A} \right\} \tag{10}$$

where $p_{v \underset{s}{\rightarrow} v'}$ denotes the conditional probability to make a transition to state v' from state v on observing symbol s.

Given the current state v_t at time t the future is conditionally independent of the past

$$
\begin{aligned}
Pr(\mathbf{s}) &= Pr\left(\mathbf{s}_t^{\leftarrow} \mathbf{s}_{t+1}^{\rightarrow}\right) \\
&= Pr\left(\mathbf{s}_{t+1}^{\rightarrow} | \mathbf{s}_t^{\leftarrow}\right) Pr\left(\mathbf{s}_t^{\leftarrow}\right) \\
&= Pr\left(\mathbf{s}_{t+1}^{\rightarrow} | v_t\right) Pr(v_t)
\end{aligned}
\tag{11}
$$

By factoring the joint distribution over the observed subsequences, the discovered states vastly reduce the dimensionality of the process's representation. In the case of strings of length N, the representation changes from requiring an N-dimensional probability vector for the joint distribution to a set of $\|\mathbf{E}\|$ transition probabilities.

Rather than simply looking up the probability of a given sequence $\mathbf{s}^N = s_0 s_1 s_2 \ldots s_{N-1}$, $s_i \in \mathbf{A}$, in the table of joint probabilities, its probability is recovered from the machine by computing the telescoping product of conditional transition probabilities

$$Pr\left(\mathbf{s}^N\right) = p_{v_0} p_{v_0 \underset{s_0}{\rightarrow} v_1} p_{v_1 \underset{s_1}{\rightarrow} v_2} \cdots p_{v_{N-1} \underset{s_{N-1}}{\rightarrow} v_N} \tag{12}$$

Here v_0 is the unique start state. By the nature of machine reconstruction it is the state of total ignorance: before the first measurement is made $p_{v_0} = 1$. The sequence $v_0, v_1, v_2, \ldots, v_{N-1}, v_N$ are those states through which the sequence s^N drives the machine.

Recall Chomsky's criterion for a scientific theory quoted at the beginning. A reconstructed machine is, in his sense, a theory of the source of the data from which it was inferred. In addition, a machine presumes to ascribe structure to unobserved sequences, such as whether or not they occur and their probabilities, by the above telescoping product (12). In fact, a machine typically does so for an indefinite number of unseen measurements. And this too is in accord with Chomsky's remark that a theory "predict an indefinite number of new phenomena".

A machine is a compact and very informative representation of the underlying source. In order to appreciate the properties that it captures, there are several statistics that can be computed from a given machine. Each of these is, naturally, a simplification in one way or another of the structure captured by the machine. Up to this point there has been no restriction on the number $\|V\|$ of states. For the sake of simplicity in the following the number of states will be assumed to be finite. This restricts the general model class to that of stochastic finite automata.

One useful reduction of a machine M is to ask for its equivalent Markov process. This is described by the stochastic connection matrix

$$T = \sum_{s \in A} T^{(s)} \tag{13}$$

where $(T)_{vv'} = p_{v \to v'}$ is the state to state transition probability, unconditioned by the measurement symbols. By construction every state has an outgoing transition. This is reflected in the fact that T is a stochastic matrix: $\sum_{v' \in V} p_{vv'} = 1$. It should be clear from dropping the input-alphabet transition labels from the machine that the detailed, I call it "computational", structure of the input data stream has been lost. All that is retained in T is the state transition structure and this is a Markov chain. The interesting fact is that Markov chains are a proper subset of stochastic finite machines. Examples latter on will support this contention. But it is exactly at this step of unlabeling the machine that the "properness" relation between these two model classes appears.

The stationary state probabilities $\vec{p}_V = \left\{ p_v : \sum_{v \in V} p_v = 1, v \in V \right\}$ are given by the left eigenvector of T

$$\vec{p}_V T = \vec{p}_V \tag{14}$$

The entropy rate of the Markov chain is then[12]

$$h_\mu(T) = - \sum_{v \in V} p_v \sum_{v' \in V} p_{v \to v'} \log_2 p_{v \to v'} \tag{15}$$

As it is an average of transition uncertainty over all the states, it measures the information production rate in bits per time step. It is also the growth rate of the Shannon information in subsequences

$$h_\mu = \lim_{L \to \infty} \frac{H(L)}{L} \tag{16}$$

where $H(L) = - \sum_{\omega \in \{s^L\}} Pr(\omega) \log_2 Pr(\omega)$. That is, $H(L) \underset{L \to \infty}{\propto} h_\mu L$. In general, subsequences are not in a one-to-one correspondence with the Markov chain's state-to-state transition sequences. Nonetheless, it is a finite-to-one relationship. And so, the Markov entropy rate is also the entropy rate of the original data source: $h_\mu(M) = h_\mu(T)$. More directly, this is given by

$$h_\mu(M) = -\sum_{v \in \mathbf{V}} p_v \sum_{v' \in \mathbf{V}} \sum_{s \in \mathbf{A}} p_{v \to v'}^{s} \log_2 p_{v \to v'}^{s} \tag{17}$$

Thus, once a machine is reconstructed from a data stream, its entropy is an estimate of underlying process's entropy rate.

The complexity[*] quantifies the information in the state-alphabet sequences

$$C_\mu(M) = H(\vec{p}_\mathbf{V}) = -\sum_{v \in \mathbf{V}} p_v \log_2 p_v \tag{18}$$

It measures the amount of memory in the source. For completeness, note that there is an edge-complexity that is the information contained in the asymptotic edge distribution $\vec{p}_\mathbf{E} = \left(p_e = p_v p_{v \to v'}^{s} : e \in \mathbf{E} \right)$

$$C_\mu^e(M) = -\sum_{e \in \mathbf{E}} p_e \log_2 p_e \tag{19}$$

These quantities are not independent. Conservation of information leads to the relation

$$C_\mu^e = C_\mu + h_\mu \tag{20}$$

Thus, there are only two independent quantities when modeling a source as a stochastic finite automaton. The entropy h_μ, as a measure of the diversity of patterns, and the complexity C_μ, as a measure of memory, have been taken as the two elementary "information processing" coordinates with which to analyze a range of sources.[19]

There is another set of quantities that derive from the skeletal structure of the machine. If we drop all probabilistic structure on the machine, the growth rate of the raw number of sequences it produces is the topological entropy

$$h = \log_2 \lambda(T_0) \tag{21}$$

where $T_0 = \sum_{s \in \mathbf{A}} T_0^{(s)}$ is called the connection matrix and $\lambda(T_0)$ is its principal eigenvalue. It is formed from the symbol matrices

$$\left\{ T_0^{(s)} : \left(T_0^{(s)} \right)_{vv'} = \begin{cases} 1 & p_{v \to v'}^{s} > 0 \\ 0 & \text{otherwise} \end{cases} s \in \mathbf{A} \right\} \tag{22}$$

The state and transition topological complexities are

$$C = \log_2 \|\mathbf{V}\|$$
$$C^e = \log_2 \|\mathbf{E}\| \tag{23}$$

Although it falls somewhat outside of the present discussion, it is worthwhile noting that these entropies and complexities can be integrated into a single parametrized framework. The resulting formulation gives a thermodynamics of ϵ-machines.[26]

---

[*] Within the reconstruction hierarchy this is actually the finitary complexity, since discussion is restricted to processes with a finite number of states. Although, I have not introduced this restriction in unnecessary places. Related forms of the finitary complexity have been considered before, outside of the context of reconstruction, assuming generating partitions and known equations of motion.[22–25]

2.4 Complexity

It is useful at this stage to stop and reflect on some properties of the models whose reconstruction and structure have just been described. Consider two extreme data sources. The first, highly predictable, produces a stream of 1s; the second, highly unpredictable, is an ideal random source of binary symbols. The parse tree of the predictable source is a single path of 1s. And there is a single distinct subtree, at any depth. As a result the machine has a single state and a single transition on $s = 1$: a simple model of a simple source. For the ideal random source the parse tree, again to any depth, is the full binary tree. All paths appear in the parse tree since all binary subsequences are produced by the source. There is a single subtree, of any morph depth, at all parse tree depths: the full binary subtree. And the machine has a single state with two transitions: one on $s = 1$ and one on $s = 0$. The result is a simple machine, even though the source produces the widest diversity of binary sequences. Thus, these two zero entropy and maximal entropy sources have zero complexity.

A simple gedanken experiment serves to illustrate how (finitary) complexity is a measure of a machine's memory capacity. Consider two observers **A** and **B**, each with the same model M of some process. **A** is allowed to start machine M in any state and uses it to generate binary strings that are determined by the edge labels of the transitions taken. These strings are passed to observer **B** which traces their effect through its own copy of M. On average how much information about M's state can **A** communicate to **B** via the binary strings? If the machine describes (say) a period three process, e.g. it outputs strings like $101101101\ldots$, $011011011\ldots$, and $110110110\ldots$; it has $\|V\| = 3$ states. Since **A** starts M in any state, **B** can learn only the information of the process's phase in the period 3 cycle. This is $\log_2 \|V\| = 1.584\ldots$ bits of information on average about the process's state, if **A** chooses the initial states with equal probability. However, if the machine describes an ideal random binary process, by definition **A** can communicate no information to **B**, since there is no structure in the sequences to use for this purpose. This is reflected in the fact, as already noted above, that the corresponding machine has a single state and its complexity is $C_\mu(M) = \log_2 1 = 0$. In this way, a process's complexity is the amount of information that someone controlling its start state can communicate to another.

These examples serve to highlight one of the most basic properties of complexity, as I use the term.[*] Both predictable and random sources are simple in the sense that their models

---

[*] A number of authors have considered measures of complexity that have heuristically similar properties. One of the first computation theoretic notions along these lines was "logical depth".[27] It relies on obtaining a minimal universal Turing machine program. And so, as a consequence of Kolmogorov's theorem, it is uncomputable.[28] Apparently, the first constructive measure proposed, i.e. one that could be estimated from a data stream, was the excess entropy convergence rate.[29] The closely-related total excess entropy[14,30] was also suggested as a measure of information contained in a process. This was later recoined the "stored information"[31] and also the "effective measure complexity".[24] Using the kneading calculus for one-dimensional maps, algorithms were given to estimate the number of equivalent Markov states.[22] The size of deterministic finite automata describing patterns generated by cellular automata was used to measure the development

are small. Complex processes in this view have large models. In computational terms, complex processes have, as a minimum requirement, a large amount of memory as revealed by many internal states in the reconstructed machine. Most importantly, this memory is structured in particular ways that support different types of computation. The sections below on knowledge and meaning show several consequences of computational structure.

In the most general setting, I use the word "complexity" to refer to the amount of information contained in observer-resolvable equivalence classes.[17] This approach puts the burden directly on any complexity definition to explicitly state the representation employed by the observer. For processes with finite memory, the complexity is measured by the quantities labeled above by C. The general notion, i.e. without the finiteness restruction, has been referred to as the "statistical complexity" in order to distinguish it from the Chaitin-Kolmogorov complexity,[33,28] the Lempel-Ziv complexity,[34] Rissanen's stochastic complexity,[35] and others[36,37] which are all equivalent in the limit of long data streams to the process's Kolmogorov-Sinai entropy $h_\mu\left(\vec{X}\right)$. If the instrument Π_ϵ is generating and $\mu\left(\vec{X}\right)$ is absolutely continuous, these quantities are given by the entropy rate of the reconstructed ϵ-machine, i.e. (17).[38] Accordingly, I use the phrases "entropy" and "entropy rate" to refer to such quantities. They measure the diversity of sequences that a process produces. Implicit in their definitions is the restriction that the modeler must pay computationally for each random bit. Simply stated, the overarching goal is exact description of the data stream. In the modeling approach advocated here the modeler is allowed to flip a coin or to sample the heat bath to which it may be coupled. "Complexity" is reserved in my vocabulary to refer to a process's structural properties, such as amount of memory, syntactic and semantic properties, and other types of computational capacity.

The present review is not the place to comment on the wide range of alternative notions of "complexity" that have been proposed recently for nonlinear physics. The reader is referred to the comments and especially the citations in [17,19,39]. It is important to point out, however, that the notion defined here does not require (i) knowledge of the governing equations of motion nor of the process's embedding dimension, (ii) the prior existence of exact conditional probabilities, (iii) Markov or even generating partitions of the state space, (iv) continuity and differentiability of the state variables, nor (v) the existence of periodic orbits. Within the framework of machine reconstruction such restrictions are to be viewed as verifiable assumptions. They can be given explicit form in terms of ϵ-machine properties. This is an essential criterion for any modeling framework that seeks to roll back the frontier of subjectivity to expose underlying mechanisms. Without it, there is little basis for appreciating

---

of spatial complexity.[23] The present author proposed a complexity measure based on the size of the group of symmetries in an object; Los Alamos Workshop on Dimension and Entropy (1985). Finally, the diffusion rate on a hierarchical potential was shown to exhibit similar "complexity" properties.[32] There is clearly no lack of complexity measures. During the last five years yet more have been proposed, including the finitary and more general complexities considered here. All of these are seen to be similar or different according to (i) the computational model class selected, often implicitly, and (ii) whether or not the equations of motion are assumed known.

what aspects of experimental reality are masked by a given set of modeling assumptions and, more importantly, for a study of the dynamics of modeling itself.

2.5 Compressing and Decompressing Chaos

A dynamical system, or any physical process for that matter, is a communication channel.[22,40] In the case of a chaotic physical process, the associated mechanism is viewed as being driven by a heat bath or a randomness source. The dynamics of the effective channel then connects the heat bath to the process's macroscopic, observable states.

This picture is particularly explicit in the case of stochastic automata. The machine model of the source can be reinterpreted as a transducer. A transducer is a machine that not only reads symbols from some input, taking the appropriate transitions, but also outputs a symbol determined by each transition taken. In the graphical representation the edges of a transducer are labeled with both input and output symbols. The latter may come from different alphabets. For example, the input symbols could be the binary strings considered up to this point, and the output symbols could be labels for each of the states reached after making a transition.

For now, consider both alphabets to be binary. Then the strings produced by a source are generated by the following communication channel that is built from the source's reconstructed machine. Starting from the start state, any deterministic transition, i.e. a state with only a single outgoing edge, is taken and that edge's output symbol is emitted. At states with more than one outgoing edge, a symbol is read from the heat bath and that edge is taken whose input symbol matches the symbol read. On taking the transition, the edge's output symbol is emitted. This defines a transformation from (random) strings to strings containing the constraints and structure that define the original machine.

The reverse procedure can be implemented to give a compression of a string. In this situation, as the string is read, no symbol is output when the transition is deterministic. When there is a branching, the corresponding string symbol is copied to the output. This transducer thereby removes all of the (topological) regularity of the string. The result is a (partially) random string. Coupling this compression transducer with the preceding heat bath transducer gives a data transmission scheme for a string, if both transducers are built from a model of the string.

2.6 On the Importance of Being Minimal

Given a machine model of a process, the machine's complexity, or any other statistic for that matter, need not be that of the process. The inference that some model property also holds for the process generally depends on details of the reconstruction algorithm used and the attendant assumptions and restrictions it imposes. Nonetheless, something about the underlying process has been estimated by reconstructing the machine by the above method. In order to see just what this is, this section focuses on an important property of the reconstruction method and

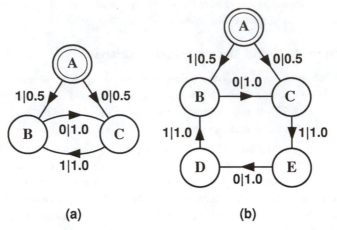

Fig. 6 Two consistent models of the period two process producing s = ... 10101010101010101 ...
(a) The minimal machine. (b) A nonminimal machine.

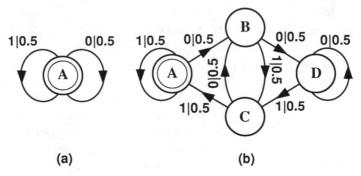

Fig. 7 Two consistent models of the ideal random process $(1 + 0)^*$. (a) The minimal machine. (b) A nonminimal machine.

the notion of state on which it is based. The property is that machine reconstruction produces the unique machine with the smallest number of states. As the following will argue, due to uniqueness and minimality there are some properties of the underlying process that can be inferred from and are well-estimated by the machine.

Figure 6 shows two machines that are (statistically) consistent models of the period two process that produces s = ... 10101010101010101 .... In fact, both machines are exact, since they describe the structure of the data stream without error.

Figure 7 shows two machines, from the other end of the entropy spectrum, that are (statistically) consistent models of the ideal random process $(1 + 0)^*$, where + means to choose s = 0 or s = 1 with equal probability. Again, both machines are exact, describing as they do the source's structure without approximation.

Minimality of the reconstructed machine ensures that its complexity is a measure, actually a lower-bound estimate, of the process's memory capacity. As just noted, the latter is the maximal amount of state-information that can be transmitted using the process as a source and allowing one to select the source's initial state. The state information that can be conveyed for a periodic process is the phase of the periodic cycle. In the present case (Fig. 6), the period was two and so one bit of phase information can be communicated. The observer receives that bit at the moment it synchronizes to the data stream.* The period two machine with five states, Fig. 6(b), has a complexity of $\log_2 5 \approx 2.321$ bits. But this amount of information clearly cannot be transmitted with the period two source. The nonminimal machine has a complexity that is too large. The case of the ideal random process is even more extreme. The nonminimal machine has a complexity of $\log_2 4 = 2$ bits, but by definition the data stream and source have no structure that can be used for communication. The minimal machine, Fig. 7(a), has zero complexity, in accord with intuition.

In summary, then, even though there is a very large number of machines consistent with a given data stream, the minimal one is singled out not only for its uniqueness and compactness, but also in order to estimate and understand the source's properties. Minimality keeps the modeler from inferring structure in the source which is not justified by the given data.

3. KNOWLEDGE RELAXATION†

The next two sections investigate how models can be used by an observer. An observer's knowledge $\mathcal{K}_P$ of a process P consists of the data stream, its current model, and how the information used to build the model was obtained.‡ Here the latter is given by the measuring instrument $\mathcal{I} = \{\Pi_\epsilon(D), \tau\}$. To facilitate interpretation and calculations, the following will assume a simple data acquisition discipline with uniform sampling interval τ and a time-independent measurement partition Π_ϵ. Further simplification comes from ignoring external factors, such as what the observer intends or needs to do with the model, by assuming that the observer's goal is optimal prediction with respect to the model class of finitary machines.

The totality of knowledge available to an observer is given by the development of its $\mathcal{K}_P$ at each moment during its history. If we make the further assumption that by some agency the observer has at each moment in its history optimally encoded the available current and past measurements into its model, then the totality of knowledge consists of four parts: the time series of measurements, the instrument by which they were obtained, and the current model and its current state. Stating these points so explicitly helps to make clear the upper bound on what the observer can know about its environment. Even if the observer is allowed arbitrary computational resources, given either finite information from a process or finite time, only a finite amount of structure can be inferred.

For the following assume that an observer's model of a process is an ϵ-machine. To see its role in the change in $\mathcal{K}_P$ consider the situation in which the model structure is kept

---

* Synchronization is the subject of a later section.
† This and the following section also appear elsewhere.[41]
‡ In principle, the observer's knowledge also consists of the reconstruction method and its various assumptions. But it is best to not elaborate this here. These and other unmentioned variables are assumed to be fixed.

fixed. Starting from the state v_0 of total ignorance about the process's state, successive steps through the machine lead to a refinement of the observer's knowledge as determined by a sequence of measurements. The average increase in $\mathcal{K}_P$ is given by a diffusion of information throughout the model. The machine transition probabilities, especially those connected with transient states, govern how the observer gains more information about the process with longer measurement sequences.

The average increase is governed by a diffusion of information throughout the given model. Consider for the moment the parse tree as a model. There is a flow of probability downwards, in increasing time, toward the leaves. This is a unidirectional diffusion of information on an ultrametric structure.[42]* The ultrametric distance on the tree is sequence length or, more simply, time itself. Taking the machine as the model, the distance between two state events, (say) A $\in$ V and B $\in$ V, is the length of the shortest directed machine path between them. The direction is determined by the order in which A and B are observed to occur. This, in turn, translates back into the difference in parse tree levels at which the associated morphs are found. When viewed in terms of either the parse tree or the machine, the data stream and its joint distribution exhibit an ultrametric structure. That structure, in turn, determines those properties of time that can be inferred from the data stream.

In a more quantitative vein, a measure of knowledge relaxation on finitary machines is given by the time-dependent finitary complexity

$$C_\mu(t) = H(\vec{p}_V(t)) \tag{24}$$

where $H(\{p_i\}) = \sum\limits_{p_i \in P} p_i \log_2 p_i$ is the Shannon entropy of the distribution $\{p_i\}$ and

$$\vec{p}_V(t+1) = \vec{p}_V(t)T \tag{25}$$

is the probability distribution at time t beginning with the initial distribution $p_V(0) = (1,0,0,\ldots)$ concentrated on the start state. This distribution represents the observer's condition of total ignorance of the process's state, i.e. before any measurements have been made, and correspondingly $C_\mu(0) = 0$. $C_\mu(t)$ is simply (the negative of) the Boltzmann H-function in the present setting. There is an analogous H-theorem for stochastic ϵ-machines: $C_\mu(t)$ converges monotonically when $\vec{p}_V(t)$ is sufficiently close to $\vec{p}_V = \vec{p}_V(\infty)$: $C_\mu(t) \underset{t\to\infty}{\to} C_\mu$. That is, the time-dependent complexity limits on the finitary complexity. Furthermore, the observer has the maximal amount of information about the process, i.e. the observer's knowledge is in equilibrium with the process, when $C_\mu(t+1) - C_\mu(t)$ vanishes for all $t > t_{\text{lock}}$, where t_{lock} is some fixed time characteristic of the process.

For finitary machines there are two convergence behaviors for $C_\mu(t)$. These are illustrated in Fig. 8 for three processes: one P_3 which is period 3 and generates $(101)^*$, one P_2 in which

---

* Unidirectional diffusion on a hierarchical structure is described by the theory of branching processes, and does not necessarily call into play the phenomena associated with ultrametric diffusion. If the direction of time is unknown, however, then the diffusion of the observer's information is bidirectional on the parse tree and so a diffusion on an ultrametric space.

only isolated zeros are allowed, and one P_1 that generates 1s in blocks of even length bounded by 0s. The first behavior type, illustrated by P_3 and P_2, is monotonic convergence from below. In fact, the asymptotic approach occurs in finite time. This is the case for periodic and recurrent Markov chains, where the latter refers to finite state stochastic processes whose support is a subshift of finite type (SSFT).[43] The convergence here is over-damped.

The second convergence type, illustrated by P_1, is only asymptotic; convergence to the asymptotic state distribution is only at infinite time. There are two subcases. The first is monotonic increasing convergence; the conventional picture of stochastic process convergence. The second subcase (P_1) is nonmonotonic convergence. Starting in the condition of total ignorance leads to a critically-damped convergence with a single overshoot of the finitary complexity. With other initial distributions oscillations, i.e. underdamped convergence, can be seen. Exact convergence is only at infinite time. This convergence type is associated with machines having cycles in the transient states or, in the classification of symbolic dynamics, with machines whose support is a strictly Sofic system (SSS).[43]* For these, at some point in time the initial distribution spreads out over more than just the recurrent states. $C_\mu(t)$ can then be larger than C_μ. Beyond this time, it converges from above. Much of the detailed convergence behavior is determined, of course, by T's full eigenvalue spectrum. The interpretation just given, though, can be directly deduced by examining the reconstructed machine's graph. One aspect which is less immediate is that for SSSs the initial distribution relaxes through an infinite number of Cantor sets in sequence space. For SSFTs there is only a finite number of Cantor sets.

This structural analysis indicates that the ratio

$$\Delta C_\mu(t) = \frac{|C_\mu - C_\mu(t)|}{C_\mu} \qquad (26)$$

is largely determined by the amount of information in the transient states. For SSSs this quantity only asymptotically vanishes since there are transient cycles in which information persists for all time, even though their probability decreases asymptotically. This leads to a general definition of (chaotic or periodic) phase and phase locking. The phase of a machine at some point in time is its current state. There are two types of phase of interest here. The first is the process's phase and the second is the observer's phase. The latter refers to the state of the observer's model having read the data stream up to some time. The observer has β-locked onto the process when $\Delta C_\mu(t_{lock}) < \beta$. This occurs at the locking time t_{lock} which is the longest time t such that $\Delta C_\mu(t) = \beta$. When the process is periodic, this notion of locking is the standard one from engineering. But it also applies to chaotic processes and corresponds to the observer knowing what state the process is in, even if the next measurement cannot be predicted exactly.

These two classes of knowledge relaxation lead to quite different consequences for an observer even though the processes considered above all have a small number of states (2 or

---

* SSS shall also refer, in context, to stochastic Sofic systems.[44]

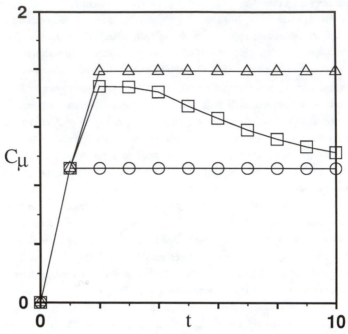

Fig. 8 Temporal convergence of the complexity $C_\mu(t)$ for a period 3 process P_3 (triangles), a Markovian process P_2 whose support is a subshift of finite type (circles), and a process P_1 that generates blocks of even numbers of 1s surrounded by 0s (squares).

3) and share the same single-symbol statistics: $Pr(s = 1) = \frac{2}{3}$ and $Pr(s = 0) = \frac{1}{3}$. In the over-damped case, the observer knows the state of the underlying process with certainty after a finite time. In the critically-damped situation, however, the observer has only approximate knowledge for all times. For example, setting $\beta = 1\%$ leads to locking times shown in table 1. Thus, the ability of an observer to infer the state depends crucially on the process's computational structure, viz. whether its topological machine is a SSFT or a SSS. The presence of extrinsic noise and observational noise modify these conclusions systematically.

It is worthwhile to contrast the machine model of P_1 with a model based on histograms, or look-up tables, of the same process. Both models are given sufficient storage to exactly represent the length 3 sequence probability distribution. They are then used for predictions on length 4 sequences. The histogram model will store the probabilities for each length 3 sequence. This requires 8 bins each containing an 8 bit approximation of a rational number: 3 bits for the numerator and 5 for the denominator. The total is 67 bits which includes an indicator for the most recent length 3 sequence. The machine model of Fig. 9 must store the current state and five approximate rational numbers, the transition probabilities, using 3 bits each: one for the numerator and two for the denominator. This gives a model size of 17 bits.

Two observers, each given one or the other model, are presented with the sequence 101. What do they predict for the event that the fourth symbol is $s = 1$? The histogram model

Locking Times at 1% Level	
Process	**Locked at time**
Period 3	2
Isolated 0s	1
Even 1 blocks	17

Table 1 β-locking times for the periodic P_3, isolated 0s P_2, and even 1s P_1, processes. Note that for the latter the locking time is substantially longer and depends on β. For the former two, the locking times indicate the times at which asymptotic convergence has been achieved. The observer knows the state of the underlying process with certainty at those locking times. For P_1, however, at $t = 17$ the observer is partially phase-locked with knowledge of 99% of the process's state information.

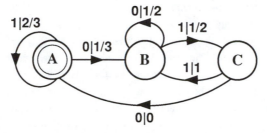

Fig. 9 The even system generates sequences $\{\ldots 01^{2n}0\ldots : n = 0, 1, 2, \ldots\}$ of 1s of even length, i.e. even parity. There are three states $\mathbf{V} = \{A, B, C\}$. The state A with the inscribed circle is the start state v_0. The edges are labeled $s|p$ where $s \in \mathbf{A}$ is a measurement symbol and $p \in [0, 1]$ is a conditional transition probability.

predicts

$$Pr(1|101) \approx Pr(1|01) = \frac{Pr(011)}{Pr(11)} = \frac{1/6}{4/9} = \frac{3}{8} \tag{27}$$

whereas the machine model predicts

$$Pr(1|101) = p_{C \to B} = 1 \tag{28}$$

The histogram model gives the wrong prediction. It says that the fourth symbol is uncertain when it is completely predictable. A similar analysis for the prediction of measuring $s = 1$ having observed 011 shows the opposite. The histogram model predicts $s = 1$ is more likely, $p_{s=1} = 2/3$, when it is, in fact, not predictable at all, $p_{s=1} = 1/2$. This example is illustrative of the superiority of stochastic machine models over histogram and similar look-up table models of time-dependent processes. Indeed, there are processes with finite memory for which no finite-size sequence histogram will give correct predictions.

In order to make the physical relevance of SSSs and their slow convergence more plausible, the next example is taken from the Logistic map at a Misiurewicz parameter value. The Logistic map is an iterated mapping of the unit interval

$$x_{n+1} = f_r(x_n) = rx_n(1 - x_n), \quad r \in [0, 4], x_0 \in [0, 1] \tag{29}$$

The control parameter r governs the degree of nonlinearity. At a Misiurewicz parameter value the chaotic behavior is governed by an absolutely continuous invariant measure. The

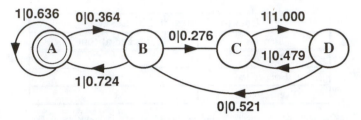

Fig. 10 The machine $M_{r'}^{\rightarrow}$ reconstructed by parsing in forward presentation order a binary sequence produced using a generating partition of the Logistic map at a Misiurewicz parameter value.

consequence is that the statistical properties are particularly well-behaved. These parameter values are determined by the condition that the iterates $f^n(x_c)$ of the map's maximum $x_c = 1/2$ are asymptotically periodic. The Misiurewicz parameter value r' of interest here is the first root of $f_{r'}^4(x_c) = f_{r'}^5(x_c)$ below that at $r = 4$. Solving numerically yields $r' \approx 3.9277370017867516$. The symbolic dynamics is produced from the measurement partition $\Pi_{1/2} = \{[0, x_c], (x_c, 1]\}$. Since this partition is generating the resulting binary sequences completely capture the statistical properties of the map. In other words, there is a one-to-one mapping between infinite binary sequences and almost all points on the attractor.

Reconstructing the machine from one very long binary sequence in the direction in which the symbols are produced gives the four state machine $M_{r'}^{\rightarrow}$ shown in Fig. 10. The stochastic connection matrix is

$$T = \begin{pmatrix} 0.636 & 0.364 & 0.000 & 0.000 \\ 0.724 & 0.000 & 0.276 & 0.000 \\ 0.000 & 0.000 & 0.000 & 1.000 \\ 0.000 & 0.521 & 0.479 & 0.000 \end{pmatrix} \tag{30}$$

Reconstructing the machine from the same binary sequence in the opposite direction gives the reverse-time machine $M_{r'}^{\leftarrow}$ shown in Fig. 11. It connection matrix is

$$T = \begin{pmatrix} 0.636 & 0.364 & 0.000 & 0.000 \\ 0.000 & 0.000 & 0.276 & 0.724 \\ 0.000 & 0.000 & 0.000 & 1.000 \\ 0.000 & 1.000 & 0.000 & 0.000 \end{pmatrix} \tag{31}$$

Notice that $M_{r'}^{\leftarrow}$ has a transient state and three recurrent states compared to the four recurrent states in $M_{r'}^{\rightarrow}$. This suggests the likelihood of some difference in complexity convergence. Figure 12 shows that this is the case by plotting $C_\mu(M_{r'}^{\rightarrow}, t)$ and $C_\mu(M_{r'}^{\leftarrow}, t)$ for positive and "negative" times, respectively. Not only do the convergence behaviors differ in type, but also in the asymptotic values of the complexities: $C_\mu(M_{r'}^{\rightarrow}) \approx 1.77$ bits and $C_\mu(M_{r'}^{\leftarrow}) \approx 1.41$ bits. This occurs despite the fact that the entropies must be and are the same for both machines: $h(M_{r'}^{\rightarrow}) = h(M_{r'}^{\leftarrow}) \approx 0.82$ bits per time unit and $h_\mu(M_{r'}^{\rightarrow}) = h_\mu(M_{r'}^{\leftarrow}) \approx 0.81$ bits per time unit. Although the data stream is equally

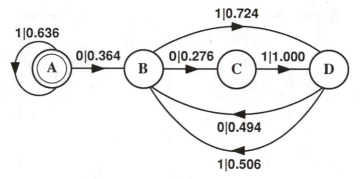

Fig. 11 The machine $M_{r'}^{\leftarrow}$ reconstructed by parsing in reverse presentation order a binary sequence produced using a generating partition of the Logistic map at a Misiurewicz parameter value.

unpredictable in both time directions, an observer learns about the process's state in two different ways and obtains different amounts of state information. The difference

$$\Delta C_{\leftarrow}^{\rightarrow} = C_\mu(M_{r'}^{\rightarrow}) - C_\mu(M_{r'}^{\leftarrow}) \approx 0.36 \text{ bits} \tag{32}$$

is a measure of the computational irreversibility of the process. It indicates the process is not symmetric in time from the observer's viewpoint. This example serves to distinguish machine reconstruction and the derived quantifiers, such as complexity, from the subsequence-based measures, such as the two-point mutual information and the excess entropy.

4. MEASUREMENT SEMANTICS

Shannon's communication theory tells one how much information a measurement gives. But what is the meaning of a particular measurement? Sufficient structure has been developed up to this point to introduce a quantitative definition of an observation's meaning. Meaning, as will be seen, is intimately connected with hierarchical representation.* The following, though, concerns meaning as it arises when crossing a single change in representation and not in the entire hierarchy.[11]

A universe consisting of an observer and a thing observed has a natural semantics. The semantics describes the coupling that occurs during measurement. The attendant meaning derives from the dual interpretation of the information transferred at that time. As already emphasized, the measurement is, first, an indirect representation of the underlying process's state and, second, information that updates the observer's knowledge. The semantic information processing that occurs during a measurement thus turns on the relationship between two levels of representation of the same event.

The meaning of a message, of course, depends on the context in which its information is made available. If the context is inappropriate, the observation will have no basis with

---

* The following hopefully adds some specificity to approaches to the symbol grounding problem: How to physical states of nature take on semantic content?[45]

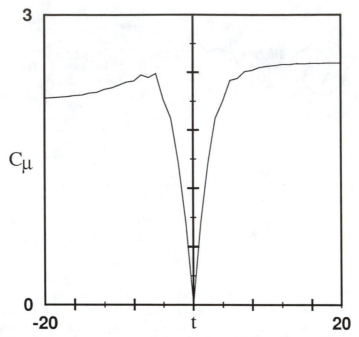

Fig. 12 What the observer sees, on average, in forward and reverse lag time in terms of the complexity convergence $C_\mu(t)$ for $M_{r'}^{\rightarrow}$ and $M_{r'}^{\leftarrow}$. Data for the latter are plotted on the negative lag time axis. Note that not only do the convergence characteristics differ between the two time directions, but the asymptotic complexity values are not equal.

which to be understood. It will have no meaning. If appropriate, then the observation will be "understood". And if that which is understood — the content of the message — is largely unanticipated then the observation will be more significant than a highly likely, "obvious" message.

In the present framework context is set by the model held by the observer at the time of a measurement. To take an example, assume that the observer is capable of modeling using the class of stochastic finite automata. And, in particular, assume the observer has estimated a stochastic finite automaton* and has been following the process sufficiently long to know the current state with certainty. Then at a given time the observer measures symbol $s \in \mathbf{A}$. If that measurement forces a disallowed transition, then it has no meaning other than that it lies outside of the contexts (morphs) captured in the current model. The observer clearly does not know what the process is doing. Indeed, formally the response is for the observer to reset the machine to the initial state of total ignorance. If, however, the measurement is associated

---

* Assume also that the estimated machine is deterministic in the sense of automata theory: the transitions from each state are uniquely labeled: $p_e = p(v, v', s) = p_v p_{v \rightarrow v'}$. This simplifies the discussion by avoiding the need to define the graph indeterminacy as a quantitative measure of ambiguity.[17] Ambiguity for an observer arises if its model is a stochastic nondeterministic automaton.

with an allowed transition, i.e. it is anticipated, then the degree $\Theta(s)$ of meaning is

$$\Theta(s) = -\log_s p_{\to v} \tag{33}$$

Here $\underset{s}{\to} v$ denotes the machine state $v \in \mathbf{V}$ to which the measurement brings the observer's knowledge of the process's state. $p_{\to v}$ is the corresponding morph's probability which is given by the associated state's asymptotic probability. The meaning itself, i.e. the content of the observation, is the particular morph to which the model's updated state corresponds. In this view a measurement selects a particular pattern from a palette of morphs. The measurement's meaning is the selected morph[*] and the degree of meaning is determined by the latter's probability.

To clarify these notions, let's consider as an example a source that produces infinite binary sequences for the regular language[48] described by the expression $(0 + 11)^*$. We assume further that the choice implied by the "+" is made with uniform probability. An observer given an infinite sequence of this type reconstructs the stochastic finite machine shown in Fig. 9. The observer has discovered three morphs: the states $\mathbf{V} = \{A, B, C\}$. But what is the meaning of each morph? First, consider the recurrent states B and C. State B is associated with having seen an even number of 1's following a 0; C with having seen an odd number. The meaning of B is "even" and C is "odd". Together the pair $\{B, C\}$ recognize a type of parity in the data stream. The machine as a whole accepts strings whose substrings of the form $01 \ldots 11 \ldots 10$ have even parity of 1s. What is the meaning of state A? As long as the observer's knowledge of the process's state remains in state A, there has been some number of 1's whose parity is unknown, since a 0 must be seen to force the transition to the parity state B. State A, a transient, serves to synchronize the recurrent states with the data stream. This indicates for this example the meaning content of an individual measurement in terms of the state to which it and its predecessors bring the machine.

Before giving a quantitative analysis the time dependence of the state probabilities must be calculated. Recall that the state probabilities are updated via the stochastic connection matrix

$$\vec{p}_{\mathbf{V}}(t+1) = \vec{p}_{\mathbf{V}}(t) \begin{pmatrix} \frac{2}{3} & \frac{1}{3} & 0 \\ 0 & \frac{1}{2} & \frac{1}{2} \\ 0 & 1 & 0 \end{pmatrix} \tag{34}$$

where $\vec{p}_{\mathbf{V}}(t) = (p_A(t), p_B(t), p_C(t))$ and the initial distribution is $\vec{p}_{\mathbf{V}}(0) = (1, 0, 0)$. Using the z-transform, the time-dependent state probabilities are found to be

$$p_A(t) = \left(\frac{2}{3}\right)^t \qquad t = 0, 1, 2, \ldots$$

$$p_B(t) = 2\left(\frac{2}{3}\right)^t - 2^{1-t} \qquad t = 0, 1, 2, \ldots$$

---

[*] I simplify here. The best formal representation of meaning at present uses the set-theoretic structure that the machine induces over the set of observed subsequences. This in turn is formulated via the lattice theory[46] of machines.[47]

$$p_C(t) = \begin{cases} \left(\frac{2}{3}\right)^t - 2^{1-t} & t = 1, 2, 3, \ldots \\ 0, & t = 0 \end{cases} \tag{35}$$

Any time a disallowed transition is forced the current state is reset to the start state and $\vec{p}_V(t)$ is reset to the distribution representing total ignorance which is given by $\vec{p}_V(0)$.

What then is the quantitative degree of meaning of particular measurements? Let's consider all of the possibilities: all possible contexts, i.e. current states, and all possible measurements. t steps after a reset, the observer is

1. In the sync state and measures $s = 1$: $\Theta_{\text{sync}}^t(1) = -\log_2 p_{\underset{1}{\to} A} = t(\log_2 3 - 1)$;

2. In the sync state and measures $s = 0$: $\Theta_{\text{sync}}^t(0) = -\log_2 p_{\underset{0}{\to} B} = -\log_2 p_B(t)$; e.g. $\Theta_{\text{sync}}^1(0) = \log_2 3 \approx 1.584$ bits;

3. In the even state and measures $s = 1$: $\Theta_{\text{even}}^t(1) = -\log_2 p_{\underset{1}{\to} C} = -\log_2 p_C(t), t > 1$; e.g. $\Theta_{\text{even}}^2(1) = \log_2 6 \approx 2.584$ bits;

4. In the even state and measures $s = 0$: $\Theta_{\text{even}}^t(0) = -\log_2 p_{\underset{0}{\to} B} = -\log_2 p_B(t)$; e.g. $\Theta_{\text{even}}^2(0) = 1 + 2\log_2 3 - \log_2 7 \approx 1.372$ bits;

5. In the odd state and measures $s = 1$: $\Theta_{\text{odd}}^t(1) = -\log_2 p_{\underset{1}{\to} B} = -\log_2 p_B(t)$; e.g. $\Theta_{\text{odd}}^3(1) = 2 + 3\log_2 3 - \log_2 37 \approx 1.545$ bits;

6. In the odd state and measures $s = 0$, a disallowed transition. The observer resets the machine: $\Theta_{\text{odd}}^t(0) = -\log_2 p_{\underset{0}{\to} A} = -\log_2 p_A(0) = 0$.

In this scheme states B and C cannot be visited at time $t = 0$ nor state C at time $t = 1$.

Assuming no disallowed transitions have been observed, at infinite time $\vec{p}_V = \left(0, \frac{2}{3}, \frac{1}{3}\right)$ and the degrees of meaning are, if the observer is

1. In the sync state and measures $s = 1$: $\Theta_{\text{sync}}(1) = -\log_2 p_{\underset{1}{\to} A} = \infty$;

2. In the sync state and measures $s = 0$: $\Theta_{\text{sync}}(0) = -\log_2 p_{\underset{0}{\to} B} = \log_2 3 - 1 \approx 0.584$ bits;

3. In the even state and measures $s = 1$: $\Theta_{\text{even}}(1) = -\log_2 p_{\underset{1}{\to} C} = \log_2 3 \approx 1.584$ bits;

4. In the even state and measures $s = 0$: $\Theta_{\text{even}}(0) = -\log_2 p_{\underset{0}{\to} B} = \log_2 3 - 1 \approx 0.584$ bits;

5. In the odd state and measures $s = 1$: $\Theta_{\text{odd}}(1) = -\log_2 p_{\underset{1}{\to} B} = \log_2 3 - 1 \approx 0.584$ bits;

6. In the odd state and measures $s = 0$, a disallowed transition. The observer resets the machine: $\Theta_{\text{odd}}(0) = -\log_2 p_{\underset{0}{\to} A} = -\log_2 p_A(0) = 0$.

Table 2 summarizes this analysis for infinite time. It also includes the amount of information gained in making the specified measurement. This is given simply by the negative binary logarithm of the associated transition probability.

Similar definitions of meaning can be developed between any two levels in a reconstruction hierarchy. The example just given concerns the semantics between the measurement symbol level and the stochastic finite automaton level.[11] Meaning appears whenever there is a change in representation of events. And if there is no change, e.g. a measurement is considered only with respect to the population of other measurements, an important special case arises.

Observer's Semantic Analysis of Parity Source				
Observer in State	Measures Symbol	Interprets Meaning as	Degree of Meaning (bits)	Amount of Information (bits)
A	1	Unsynchronized	Infinity	0.585
A	0	Synchronize	0.585	1.585
B	1	Odd number of 1s	1.585	1
B	0	Even number of 1s	0.585	1
C	1	Even number of 1s	0.585	0
C	0	Confusion: lose sync, reset to start state	0	Infinity

Table 2 The observer's semantics for measuring the parity process of Fig. 9.

In this view Shannon information concerns degenerate meaning: that obtained within the same representation class. Consider the information of events in some set E of possibilities whose occurrence is governed by arbitrary probability distributions $\{P, Q, \ldots\}$. Assume that no further structural qualifications of this representation class are made. Then the Shannon self-information $-\log p_e$, $p_e \in P$, gives the degree of meaning $-\log_2 p_{\to e}$ in the observed event e with respect to total ignorance. Similarly, the information gain $I(P; Q) = \sum_{e \in E} p_e \log_2 \frac{p_e}{q_e}$ gives the average degree of "meaning" between two distributions. The two representation levels are degenerate: both are the events themselves. Thus, Shannon information gives the degree of meaning of an event with respect to the set E of events and not with respect to an observer's internal model; unless, of course, that model is taken to be the collection of events as in a histogram or look-up table. Although this might seem like vacuous re-interpretation, it is essential that general meaning have this as a degenerate case.

The main components of meaning, as defined above should be emphasized. First, like information it can be quantified. Second, conventional uses of Shannon information are a natural special case. And third, it derives fundamentally from the relationship *across* levels of abstraction. A given message has different connotations depending on an observer's model and the most general constraint is the model's level in a reconstruction hierarchy. When model reconstruction is considered to be a time-dependent process that moves up a hierarchy, then the present discussion suggests a concrete approach to investigating adaptive meaning in evolutionary systems: emergent semantics.

In the parity example above I explicitly said what a state and a measurement "meant". Parity, as such, is a human linguistic and mathematical convention, which has a compelling naturalness due largely to its simplicity. A low level organism, though, need not have such a literary interpretation of its stimuli. Meaning of (say) its model's states, when the state sequence is seen as the output of a preprocessor,[*] derives from the functionality given to the organism, as a whole and as a part of its environment and its evolutionary and developmental history. Said this way, absolute meaning in nature is quite a complicated and contingent concept. Absolute meaning derives from the global structure developed over space and through time. Nonetheless, the analysis given above captures the representation level-to-level origin of "local" meaning. The tension between global and local entities is not the least bit new to nonlinear dynamics. Indeed, much of the latter's subtlety is a consequence of their inequivalence. Analogous insights are sure to follow from the semantic analysis of large hierarchical processes.

5. CONCLUDING REMARKS

By way of summarizing the preceding discussion, there are a few points that can be brought out concerning what reconstructed machines represent. First, by the definition of future-equivalent states, they give the minimal information dependency between the morphs. In this respect, they represent the causality of the morphs considered as events. If state B follows state A then A is a cause of B and B is one effect of A. Second, the machines capture the information flow within the given data stream. Machine reconstruction produces minimal models up to the given approximation level; that is, up to the amount of data available for the estimation and up to the setting of the parameters (D, L, δ). This minimality guarantees that there are no other events (morphs) that intervene between successive states, at the given error level, to render A and B independent. In this and only this case, can one unambiguously say that information flows from A to B, under the chosen parsing direction. The amount of information that flows is given by the mutual information $I(A; B) = H(B) - H(B|A)$ of observing the state-event A followed by state-event B. This criterion for information flow also extends to spatial systems. Finally, time is the natural ordering captured by machines. As noted in the section on knowledge relaxation changing the direction of parsing the data stream leads to a quantitative measure of a new type of irreversibility. This type of irreversibility is a consequence of the computational, in fact, semi-group, structure of the underlying process.

Causation appears then as an efficient organization of knowledge. It is a symmetry in the broadest sense of the word. It allows for the "factoring" of experience into lower dimensional representations. The conditional independence basis for machine reconstruction of states implies that ϵ-machines are the minimal causal representations of the source, up to the given approximation level. The hierarchical reconstruction procedure uses a slight, but significant extension of this in order to address the question of change model classes: An ϵ-machine for a process is the minimal causal representation reconstructed using the least powerful computational model class that yields a finite complexity.

At first glance the process of measurement, the elemental act of information acquisition, would seem to be an antecedent of causality. This impression is, though, a result of

---

[*] This preprocessor is a transducer version of the model that takes the input symbols and outputs strings in the state alphabet V.

current nonphysical formulations of information theory. The essential physical contact during measurement calls into play the entire panoply of modeling and semantics laid out above. Information theory, as an engineering discipline, only considers the amount of information and rates of its production, loss, and transmission, as measured by various entropies. As such it ignores, as Shannon said it should, the question posed in the previous Woodward proceedings when considering the geometric structure of the space of information sources and the algebra of measurements: "what of the 'meaning' of ... information?"[49] The answer here, simply stated, is that it depends on the current state of knowledge of the observer. In restricting the class of models to stochastic finitary automata, a concrete answer was given in terms of the discovered morphs and the observer's anticipation of seeing them. A more general answer, though, lies in clearly delineating the resources, computational and observational, available to the modeler. Lending this large a context to a quantitative theory of meaning, though, pushes the boundaries especially of theoretical computer science, from which we must know how space and time complexities trade-off against one another. At present, these scalings appear to be largely, if not exclusively, studied as separate coordinates for the space of computational tasks.[48,50] Future progress demands a more detailed understanding of the complexity-entropy landscape over the space of computational tasks. There are some provocative hints from phenomenological studies that phase transitions organize some aspects of this space.[19,51,52]

What is the role of nonlinearity in all of this? I would claim at this point that it is much more fundamental than simply providing an additional and more difficult exercise in building good models and formalizing what is seen. Rather it goes to the very heart of genuine discovery. Let me emphasize this by way of a contrapositive point.

Linear systems are meaningless in the sense that the question of semantics need not arise within their universe of discourse. The implementations, artificial or natural, of mechanisms by which a linear system could be inferred are themselves nonlinear. Said another way there is no finite linear system that learns stably. Therefore, (finite) linear systems as a class are not self-describing; the universe is open. This indirect observation suggests that there is an intimate connection between meaning and nonlinearity, since nontrivial semantics requires the ability to model.

The utility of complexity, when it is seen as the arbiter of order and randomness, within nonlinear physics can be illustrated by way of posing some final questions. Consider the evolution of scientific theories. In particular, focus on the theories of time-dependent physical processes, viz. Laplacian-Newtonian classical mechanics (LNCM) and Copenhagen quantum mechanics (CQM). In LNCM (local) determinism implies complete (local) predictability. In CQM (local) nondeterminism implies complete (local) unpredictability and so only statistical regularity can be present. Shrödinger's equation is, in fact, the equation of motion governing that statistical regularity. In light of the preceding discussion of modeling, it is rather curious, but probably no coincidence, that these two theories come from the two extremes of entropy.

The meaning of physical events is often couched only in terms of these two contenders via (say) Bohr's notion of complementarity. But what about the intermediate possibility: complexity? Could these two opposed views be incomplete aspects, or projections, of a complex nature? A nature too intricate and too detailed to be completely understood at any one time and with finite knowledge? Whose state is too complex to determine with any finite measurement?

It seems that such issues will only be given their proper framing when we understand how physical nature intrinsically computes and how subsystems would spontaneously take up the task of modeling. Then, of course, there is the nonlinear dynamics of this process. Could this natural complexity be a state to which a system evolves spontaneously? A true self-organized complexity?

Hopefully, the preceding made it clear that to ignore the central role of computation is to miss the point of these questions entirely.

Many thanks to the Santa Fe Institute, where the author was supported by a Robert Maxwell Foundation Visiting Professorship, for the warm hospitality during the writing of the present essay. Funds from NASA-Ames University Interchange NCA2-488 and the AFOSR also contributed to this work.

REFERENCES

1. N. Chomsky, "Three models for the description of language," IRE Trans. Info. Th. **2**, 113 (1956).

2. A. O'Hear, **An Introduction to the Philosophy of Science** (Oxford University Press, Oxford, 1989).

3. D. M. MacKay, **Information, Meaning and Mechanism** (MIT Press, Cambridge, 1969).

4. S. Omohundro, "Modelling cellular automata with partial differential equations," Physica **10D**, 128 (1984).

5. J. P. Crutchfield, "Turing dynamical systems," preprint (1987).

6. L. Blum, M. Shub, and S. Smale, "On a theory of computation over the real numbers," Bull. AMS **21**, 1 (1989).

7. C. Moore, "Upredictability and undecidability in dynamical systems," Phys. Rev. Lett. **64**, 2354 (1990).

8. D. C. Dennett, **The Intentional Stance** (MIT Press, Cambridge, 1987).

9. J. P. Crutchfield and B. S. McNamara, "Equations of motion from a data series," Complex Systems **1**, 417 (1987).

10. J. Rissanen, **Stochastic Complexity in Statistical Inquiry** (World Scientific, Singapore, 1989).

11. J. P. Crutchfield, "Reconstructing language hierarchies," in **Information Dynamics**, ed., H. A. Atmanspracher (Plenum, New York, 1990).

12. C. E. Shannon and W. Weaver, **The Mathematical Theory of Communication** (University of Illinois Press, Champaign-Urbana, 1962).

13. A. N. Kolmogorov, "A new metric invariant of transient dynamical systems and automorphisms in lebesgue spaces," Dokl. Akad. Nauk. SSSR **119**, 861 (1958). (Russian) Math. Rev. vol. 21, no. 2035a.

14. J. P. Crutchfield and N. H. Packard, "Symbolic dynamics of noisy chaos," Physica **7D**, 201 (1983).

15. N. H. Packard, J. P. Crutchfield, J. D. Farmer, and R. S. Shaw, "Geometry from a time series," Phys. Rev. Let. **45**, 712 (1980).

16. H. Poincare, **Science and Hypothesis** (Dover Publications, New York, 1952).

17. J. P. Crutchfield and K. Young, "Inferring statistical complexity," Phys. Rev. Let. **63**, 105 (1989).

18. J. Pearl, **Probabilistic Reasoning in Intelligent Systems** (Morgan Kaufman, New York, 1988).

19. J. P. Crutchfield and K. Young, "Computation at the onset of chaos," in **Entropy, Complexity, and the Physics of Information**, ed., W. Zurek, **VIII of SFI Studies in the Sciences of Complexity** (Addison-Wesley, Reading, Massachusetts, 1990) 223.

20. A. Fraser, "Using hidden markov models to predict chaos," preprint (1990).

21. L. R. Rabiner, "A tutorial on hidden markov models and selected applications," IEEE Proc. **77**, 257 (1989).

22. J. P. Crutchfield, **Noisy Chaos**. PhD thesis, University of California, Santa Cruz (1983). Published by University Microfilms Intl, Minnesota.

23. S. Wolfram, "Computation theory of cellular automata," Comm. Math. Phys. **96**, 15 (1984).

24. P. Grassberger, "Toward a quantitative theory of self-generated complexity," Intl. J. Theo. Phys. **25**, 907 (1986).

25. K. Lindgren and M. G. Nordahl, "Complexity measures and cellular automata," Complex Systems **2**, 409 (1988).

26. J. P. Crutchfield and K. Young, "ϵ-machine spectroscopy," preprint (1991).

27. C. H. Bennett, "Dissipation, information, computational complexity, and the definition of organization," in **Emerging Syntheses in the Sciences**, ed., D. Pines (Addison-Wesley, Redwood City, 1988).

28. A. N. Kolmogorov, "Three approaches to the concept of the amount of information," Prob. Info. Trans. **1**, 1 (1965).

29. J. P. Crutchfield and N. H. Packard, "Noise scaling of symbolic dynamics entropies," in **Evolution of Order and Chaos**, ed., H. Haken (Springer-Verlag, Berlin, 1982) 215.

30. N. H. Packard, **Measurements of Chaos in the Presence of Noise**. PhD thesis, University of California, Santa Cruz (1982).

31. R. Shaw, **The Dripping Faucet as a Model Chaotic System** (Aerial Press, Santa Cruz, California, 1984).

32. C. P. Bachas and B. Huberman, "Complexity and relaxation of hierarchical structures," Phys. Rev. Let. **57**, 1965 (1986).

33. G. Chaitin, "On the length of programs for computing finite binary sequences," J. ACM **13**, 145 (1966).

34. A. Lempel and J. Ziv, "On the complexity of individual sequences," IEEE Trans. Info. Th. **IT-22**, 75 (1976).

35. J. Rissanen, "Stochastic complexity and modeling," Ann. Statistics **14**, 1080 (1986).

36. W. H. Zurek, "Thermodynamic cost of computation, algorithmic complexity, and the information metric," preprint (1989).

37. J. Ziv, "Complexity and coherence of sequences," in **The Impact of Processing Techniques on Communications**, ed., J. K. Skwirzynski (Nijhoff, Dordrecht, 1985) 35.

38. A. A. Brudno, "Entropy and the complexity of the trajectories of a dynamical system," Trans. Moscow Math. Soc. **44**, 127 (1983).

39. J. P. Crutchfield, "Inferring the dynamic, quantifying physical complexity," in **Measures of Complexity and Chaos**, eds., N. B. Abraham, A. M. Albano, A. Passamante, and P. E. Rapp (Plenum Press, New York, 1990) 327.

40. R. Shaw, "Strange attractors, chaotic behavior, and information flow," Z. Naturforsh. **36a**, 80 (1981).

41. J. P. Crutchfield, "Semantics and thermodynamics," in **Nonlinear Modeling**, eds., M. Casdagli and S. Eubank, SFI Studies in the Sciences of Complexity (Addison-Wesley, Reading, Massachusetts, 1991).

42. R. Rammal, G. Toulouse, and M. A. Virasoro, "Ultrametricity for physicists," Rev. Mod. Phys. **58**, 765 (1986).

43. B. Marcus, "Sofic systems and encoding data," IEEE Transactions on Information Theory **31**, 366 (1985).

44. B. Kitchens and S. Tuncel, "Finitary measures for subshifts of finite type and sofic systems," Memoirs of the AMS **58**, no. 338 (1985).

45. S. Harnad, "The symbol grounding problem," Physica **42D**, 335 (1990).

46. G. Birkhoff, **Lattice Theory** third ed. (American Mathematical Society, Providence, 1967).

47. J. Hartmanis and R. E. Stearns, **Algebraic Structure Theory of Sequential Machines** (Prentice-Hall, Englewood Cliffs, 1966).

48. J. E. Hopcroft and J. D. Ullman, **Introduction to Automata Theory, Languages, and Computation** (Addison-Wesley, Reading, 1979).

49. J. P. Crutchfield, "Information and its metric," in **Nonlinear Structures in Physical Systems - Pattern Formation, Chaos and Waves**, eds., L. Lam and H. C. Morris (Springer-Verlag, New York, 1990).

50. M. R. Garey and D. S. Johnson, **Computers and Intractability: A Guide to the Theory of NP-Completeness** (W. H. Freeman, New York, 1979).

51. C. G. Langton, "Computation at the edge of chaos: Phase transitions and emergent computation," in **Emergent Computation**, ed., S. Forrest (North-Holland, Amsterdam, 1990) 12.

52. W. Li, N. H. Packard, and C. G. Langton, "Transition phenomena in cellular automata rule space," preprint (1990).

Complexity Issues in Robotic Machine Learning of Natural Language

P. Suppes, L. Liang and M. Böttner

In Sections 1, 2 and 3, the theoretical framework we have been developing for a probabilistic theory of machine learning of natural language is outlined. In Section 4, some simple examples showing how mean learning curves can be constructed from the theory are given. But we also show that the explicit computation of the mean learning curve for an arbitrary number of sentences is unfeasible. This result holds even when the learning itself is quite rapid. In Section 5 we briefly describe the kinds of comprehension grammars generated by our theory from a given finite sample of sentences.

1 Introduction

In order to have a framework to discuss complexity issues in the learning of natural language, we must first describe, even if intuitively and somewhat too briefly, our probabilistic theory of natural-language learning. More than most other current approaches, we have taken a very explicit route that uses principles of association and generalization derived from classical psychological principles, but as will be evident enough, the theory we develop is in no sense something that falls within the domain of classical behavior theory — if for no other reason than the extensive internal memory structure we introduce.

A second point we want to emphasize is that our theory has been developed to simultaneously learn English, German, and Chinese. For quickness of reference, what we have to say here will almost entirely be in terms of

Patrick Suppes, Lin Liang: Institute for Mathematical Studies in the Social Sciences, Stanford University, Stanford, CA 94305-4115
Michael Böttner: Max-Planck-Institute for Psycholinguistics, Nijmegen, The Netherlands. Research supported by Grant Le 683/1-2 from the Deutsche Forschungsgemeinschaft.

English, but it is important to stress that the ideas set forth have been tested as well on the learning of German or Chinese. The third general point is that we have also carried out an implementation on Robotworld, a robotic system now extensively used in experimental work in robotics in the United States.

There are two other general points about our work to be stressed. Our viewpoint toward language learning is in most respects semantical, rather than syntactical. The formation of grammatical forms, a key aspect of our theory, is driven by a uniform set of semantic categories, rather than grammatical categories, which may vary from one language to another. The final general point that is an important limitation on the theory as currently developed is that it is solely concerned with language comprehension. Nothing that we have to say in this paper will deal in a serious way with problems of language production.

The rest of this section is focused on an intuitive description of the learning theory we develop. Section 2 states in systematic fashion our learning axioms, and Section 3 outlines the structure of the internal representations we use. Section 4 develops the theory of mean learning curves, with emphasis on their complexity. Section 5 describes briefly our grammar generation process. In Section 6, the final one, we sketch the relation between our work and some similar projects.

1.1 BACKGROUND COGNITIVE AND PERCEPTUAL ASSUMPTIONS

Before formulating the learning principles we use, we need to state informally certain assumptions about the cognitive and perceptual capacities of the class of robots we work with.

Internal Language. We assume the robot has a fully developed internal language which it does not learn. It is technically important, but not conceptually fundamental that in the present case this language is LISP. (For more details, see Section 3.) What must be emphasized is that when we speak here of the *internal language* we refer only to the language of the internal representation, which is itself a language of a higher level of abstraction, relative to the concrete movements and perceptions of the robot. There are several other internal language modules that were either developed by us or that come in the software package with Robotworld. It has been important to us that most of the machine learning of a given natural language can take place through simulation of the robot's behavior just at the level of the language of the internal representation.

Objects, Relations and Properties. We furthermore assume the robot begins its natural language learning with all the basic cognitive and perceptual concepts it will have. This means that our first language learning experiments are pure language learning. For example, we have assumed that the spatial relations frequently referred to in all the languages we consider in detail are already known to the robot. This is of course contrary to human language learning. There is good evidence, for example, that most children do not understand the relations of left and right, even at the age of thirty-six months when their command of language is already extremely good.

Actions. What was just said about objects and relations applies also to actions. There are in the internal language symbols for a fixed set of possible actions. The problem is only to learn their particular linguistic representation in a given language. What has been said about objects, relations, properties and actions constitutes a permanent part of memory of the robot. This memory does not change and is therefore not involved in the learning theory we formulate. (It is obvious that in a more general theory we will want to include learning of new concepts, etc.)

1.2 INTUITIVE DESCRIPTION OF THE PROCESS OF LEARNING

What we want to describe now is the process of learning in terms of the various events that happen on a given trial. First, the robot begins a trial in a given state of memory. There are three parts of this memory that are changed due to learning. The first is the association relation between words of a given language and internal symbols that represent denotations of members of categories in the internal representation language. For example, the action of putting will be represented internally, let us say, by the symbol p and will have three different linguistic representations in English, German and Chinese. The problem will be to learn in each language what word is properly associated with the internal representation p. The same process of association must be learned for objects, properties and spatial relations. But knowing such associations is not enough, contrary to some naive theories of associationism.

It is also important to have grammatical forms. We will not try to lay out everything about grammatical forms that we have in our fully stated theory, but it will be useful to give some examples. Consider the verbal command *Get the screw!* This would be an instance of the grammatical form A *the* O!, where A is the category of actions and O is the category of objects. As might be expected we do not actually have just a single category of actions, but several subcategories, depending upon the number of arguments required, etc. The central point here, however, is the nature of a grammatical form. The grammatical forms are derived only from actual instances of verbal commands given to the robot. Secondly, associated with each grammatical form are the first associations of words with their internal interpretations. So for example, if *Get the screw!* had been the first occurrence on which the grammatical form just stated was generated, then also stored with that grammatical form would be the associations *get* $\sim$ g, *screw* $\sim$ s—we use g for the internal symbol corresponding correctly to *get*, and sometimes we just use *g* for *get*, as in the trees shown later. Similar conventions apply to other denoting words. The third part of memory that varies is the short-term memory that holds a given verbal command for the period of the trial on which it is effective. This memory content decays and is not available for access after the trial on which a particular command is given. So at the beginning of the trial this short term buffer is empty, but is filled by the second step in learning. A verbal command is given to the robot and it is held in memory in the short-term buffer.

The third step is for the learning program to look up the associations of the words in the verbal command that has been given. If associations exist for any of the words, the categories of the associations, which are the categories of the internal interpretation, are also retrieved. The categories are substituted for the associated words and an effort is then made to generate recursively the resulting grammatical form. For example, if mistakenly the word *get* had been associated to $s, the internal symbol for *screw*, and *screw* had been associated with $g, the internal symbol for *get*, then the grammatical form that would have been generated and now found upon a second occurrence of the verbal command would be *O the A*! Now if there were no such grammatical form generated, once the associations were formed such a grammatical form would be created by the process of generalization which is used to generate grammatical forms.

When a grammatical form is stored in memory, also associated with this grammatical form in memory is its internal representation. This is an important part of the memory that changes with learning as well. If the grammatical form is generated, the internal representation is then used to execute the verbal command that has been given. If the command is executed correctly then the robot is ready for a new learning trial.

The important case of learning is when no grammatical form can be generated recursively from the grammatical forms in memory to match the form of the given verbal command. In this case the robot is unable to make a response. The correct response must be coerced. On the basis of this coercion, a new internal representation is formed. At this point the critical step comes of a probabilistic association between words of the verbal utterance and the internal denoting symbols of the new internal representation. This probabilistic association is assumed to be on a uniform probability basis. For example, if within the new internal representation there are two internal denoting symbols and there are no words in the verbal command associated to either of these internal symbols, and there are four words in the verbal command, then any pair of the four will be as likely to be associated with the pair of internal symbols as any other pair. After this association is made, a new grammatical form is generated, and possibly because of the new associations at least one of the old grammatical forms is deleted, for one of the axioms states that a word or internal denoting symbol can have exactly one association, so when a new association is formed for a word any old associations must be deleted. (This strong all-or-none uniqueness assumption is undoubtedly too restrictive for actual language learning by children, but there is some evidence that it holds in the early stages of language learning. It works very well for the kind of systematic language and grammar we are concerned to use in robot discourse, at least at these early stages of development.) With the new associations formed and old ones possibly deleted, the robot is ready for the next trial in a new state of memory.

2   Learning Principles

To facilitate the statement of principles governing the learning process just described, certain notational conventions are useful. First, generally we use Latin letters to refer to verbal commands or their parts, whatever the natural language, and we use Greek letters to refer to internal representations or their parts. The letters a, a_i a'_i, etc. refer to words in a verbal command, and the Greek letters $\alpha, \alpha_j, \alpha'_j$, etc. to internal denotations. The Roman letter t, as well as t_i, t'_i refer to terms of a verbal command, and correspondingly τ, τ_i, τ'_i to terms of an internal representation, i.e., in the present setup, LISP expressions. The symbols s, s' and $s(t)$, showing that t is a term of s, refer to entire verbal commands, and correspondingly $\sigma, \sigma', \sigma(\tau)$ for entire internal representations. Grammatical forms — either sentential or term forms — are denoted by g or also $g(X)$ to show a category argument of a form; correspondingly the internal representations of a grammatical form are denoted by γ or $\gamma(X)$. We violate our Greek-Latin letter convention in the case of semantic categories or category variables X, X', Y, etc. We use the same category symbols in both grammatical forms and their internal representations. We now turn to the statement of the axioms of learning in intuitive form. We will give a more formal and explicit statement in a longer version. We have delayed until Section 5 statement of the most technical axioms, the one on term association and the one on term form substitution, which are used to generate recursive grammatical forms.

2.1   Axioms of Learning

1. (*Association by contiguity*). If verbal command s is contiguous with a coerced action that has internal representation σ, then s is associated with σ, i.e., in symbols $s \sim \sigma$.

2. (*Probabilistic association*). If $s \sim \sigma, s$ has a set $\{a_i\}$ of denoting words not associated with any internal denotations of σ, and σ has a set $\{\alpha_j\}$ of internal symbols not associated with any words of s, then an element of $\{a_i\}$ is uniformly sampled without replacement from $\{a_i\}$, at the same time an element of $\{\alpha_j\}$ is correspondingly sampled, and

the sampled pair are associated, i.e., $a_i \sim \alpha_j$. Sampling continues until there is no remaining a_i or α_j.

3. (*Prior associations*). When a word in a verbal command or an internal symbol is given a new association (Axiom 2), any prior associations of that word or symbol are deleted from memory.

4. (*Forgetting associations of commands*). An association $s \sim \sigma$ of a verbal command s is held only temporarily in memory until the action represented internally by σ is executed or until all word associations are complete (Axiom 2).

5. (*Correction procedure*). If a verbal command s cannot be executed or is executed incorrectly with $s \sim \sigma$, and if a correct response with internal representation σ' is coerced, then s is associated with σ' for application of Axiom 2.

6. (*Category generalization*). If $t \sim \tau$ and $\tau \in X$ then $t \in X$.

7. (*Grammatical form generalization*). If $g(t) \sim \gamma(\tau)$, $t \sim \tau$ and $t \in X$, then $g(X) \sim \gamma(X)$.

8. (*Term specification*). If $g(X) \sim \gamma(X)$, $t \sim \tau$ and $t \in X$, then $g(t) \sim \gamma(\tau)$.

9. (*Memory trace for a grammatical form*). The first time a grammatical generalization (X) (Axiom 7) is formed, the word associations on which the generalization is based are stored with it.

10. (*Elimination of a grammatical form*). If a memory-trace association $a \sim \alpha$ for g is eliminated (Axiom 3), then g is eliminated from memory.

3 Internal Representation

In this section we sketch the language of internal representation we use. As already mentioned, for purposes of our robotic application we use LISP as the language of representation. Internal representations will therefore be LISP-expressions. A LISP-expression is any string $(E_1...E_n)$ where $E_1, ..., E_n$ are either atoms or LISP-expressions.

The fragments of natural language are meant to instruct a robot to perform elementary actions in a simple environment. Due to the capabilities of the robot the tasks are moving around in a 3D space and opening (and closing) the gripper. The environment is a collection of objects like screws, nuts, washers, plates, sleeves of a limited number of colors, sizes, and shapes. Commands that typically arise in this context are: *Open the gripper! Move forward! Turn to the left! Put a nut on the screw! Go to the right! Go to a screw! Lift the black screw which is left of the plate! Drop the washer! Get a nut! Put the nut near the screw! Turn the screw to the right!* and *Stop!*

So we will have in our internal language symbols for each content word of a command. The English command

$$Get\ the\ screw!$$

will have the following internal representation

$$(fa1\ \$g\ (io\ (fo\ \$s\ *)))$$

where $\$g$ and $\$s$ occur in the internal language as counterparts of the content words *get* and *screw* of English. The internal representation has three more symbols: *fo, io,* and *fa1*. These symbols are abbreviations of the semantic operations *focus on object, identify object,* and *form action*. The purpose of these operations is to provide what we think is the procedural structure of the natural language command mentioned above. Before it can perform the action denoted by $\$g$, the robot has to determine the object of this action. The object that is intended to become the object of the action is returned by the operation *io*. This operation identifies one out of a set of objects. The input for *io* is a set of objects. This set of objects is the output of the operation *fo*: it takes a property and the set of objects in the robot's environment and returns a set of objects. The environment is represented by the symbol $*$ in our internal language. In our example, the property the environment is checked for is the presence of screws. The procedure *io* then takes this set of all screws of the environment as input and returns a unique screw from this set. This particular screw together with the action $\$g$ is the input of the operation *fa1*.

According to the semantic category of the object that it denotes, each expression of our language of internal representation belongs to a certain category. In the above example, e.g., the expression *screw* belongs to the category property, more specifically: object property, and the expression *get* belongs to the category action. The list of categories is as follows: Property (P), Spatial Relation, (R), Action (A), and Object (O).

Semantic categories can be further divided into subcategories. The category R of spatial relations is split up into subcategories R_1 and R_2 depending on whether its elements are binary or ternary relations. Associated with the category R is the semantic operation *form property*, abbreviated as fp, which takes a relation and object as input and returns a property — examples are to be found in Section 5. The category A of actions has six subcategories depending on the valency of the action expression. So we have a subcategory for actions that do not require any complement like *stop*, another subcategory for actions that require an object like *get*, a subcategory for actions that require a region as complement like *go* as in, e.g. *Go near a plate!* a subcategory for actions that require both an object and

a region like *put*, as in e.g. *Put a screw near a plate!* a subcategory for
actions that require a direction as their complement like *turn*, as in e.g.
Turn left!, and a subcategory for actions that require both an object and
a direction as in e.g. *Move the screw forward!* Unlike the subcategories of
relations the subcategories of actions are not disjoint, i.e. an action like,
e.g. *turn*, occurs in more than one category.

4 Mean Learning Curves

Mean learning curves represent the average of all possible individual learning curves. Such mean curves have several important features. First, by
abstracting from the details of individual curves they give a sense of the
rate of learning to be expected. In the case of machine learning this can
be important in evaluating the practicality of the theory proposed. If the
expected number of trials to reach a satisfactory learning criterion is 2^{1000},
for example, then the theory is not of practical interest. Second, mean
learning curves typically exhibit a theoretical robustness that is not characteristic of individual learning curves, i.e., in probabilistic terms, individual sample paths. Many minor theoretical details and, on occasion, major
ones as well, can be modified without changing the theoretically predicted
mean learning curve. A familiar example is the mean learning curve that
is identical for one-parameter linear incremental models of simple learning and one-parameter all-or-none Markov models of the same phenomena.
The predicted individual sample paths are completely different for the two
kinds of models, but the mean learning curves are identical.

4.1 SIMPLE EXAMPLES

Because in simple cases we can theoretically compute the mean learning
curve from the axioms of Section 2, we will first consider some very simple
cases that give an insight into how things work. These cases are so simple
that no actual simulation is needed. In the analyses we shall consider, let

m = number of distinct commands,
d = number of distinct denoting words,
k = average number of denoting words per command.

In the case of parameter k, in the simple cases we consider k is not simply
an average but a constant.

Case 1. $m = 3$, $d = 3$, $k = 1$. An example of this case would be simply
three commands: *Forward!*, *Left!* and *Right!* Because we have three commands and three denoting words with one occurring in each sentence, it is
quite easy to derive the mean probability of a correct response on trial n.
It is apparent at once under the condition that we present a block of the
three commands randomly sampling without replacement. Such a block is

then repeated under the same conditions, that is, random sampling without replacement. In the present case, it is clear that on first appearance each command is not understood by the robot and consequently there will be no response, so the probability of a correct response is zero. Secondly, once each command is given it will be learned with probability 1. In this framework it is easy to see that

$$p_1 = p_2 = p_3 = 0 \text{ and } p_n = 1,$$

for $n \geq 4$. In this case it is also obvious that the mean curves and the individual learning curves are identical because of the simplicity of the situation.

Case 2. $m = 2$, $d = 2$, $k = 2$. An example of this case would be the two commands: *Left, right!* and *Right, left!* It is easy to write down the tree structure from which we may derive the mean learning curve. Let ℓ stand for *left* and r for *right*. We use a superscript $*$ to show the probabilistic association is correct. Thus $\ell^* r^*$ means the command *Left, right!* has the correct associations for both denoting words, whereas $\underline{\ell} \underline{r}$ means both words have wrong associations. The command given at the beginning of a trial is shown in parentheses. So (ℓr) at a vertex means the trial in question begins with the command *Left, right!* In the tree (see Figure 1) we start trial 1

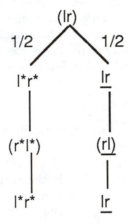

FIGURE 1. Partial Tree for Case 2. $m = d = k = 2$.

this way, followed by the two branches of possible associations, each branch having probability $\frac{1}{2}$, in accordance with Axiom 2. But in this special case, no further branching after the first trial need occur for the responses are always correct. So $p_1 = 0$ and $p_n = 1$ for $n \geq 2$. It will clarify the way the grammatical forms work to explain this result. For this purpose we examine only the right branch of the tree. On this branch on trial 1 the associations are the incorrect ones.

$$\ell \sim \$r \ ,$$

$$r \sim \$\ell \ .$$

The grammatical form generated and its associated internal form are then:

$$A' \ A \ \sim \ I(A, A')$$

Note that on the left branch this reversal of order of A and A' between the grammatical form and the associated internal form does not take place:

$$A \ A' \ \sim \ I(A, A') \qquad (left \ branch)$$

But in this restricted, simple-minded but instructive, example, either grammatical form works for its branch.

The next two cases are perhaps the simplest which have a nontrivial mean learning curve, i.e., the curve is not just a (0,1) step function.

Case 3. $m = 2, d = 3, k = 2$, *but with a special sampling procedure*. The special sampling procedure is that rather than randomizing separately each block of two commands, the two commands are simply alternated on each trial. An example of this case would be: *Get nut!* and *Get screw!*, which, following our earlier notation, we abbreviate as $g\,n$ and $g\,s$ respectively. The association tree in the first few trials has the form shown in Figure 2 (where the choice of which command to begin with clearly does not matter). So, as is easily shown on trial n

$$P_n \ (\text{Correct response}) \ = 1 - \frac{1}{2^{n-2}},$$

for $n \geq 2$, and thus $p_1 = 0$, $p_2 = 0$, $p_3 = \frac{1}{2}$, $p_4 = \frac{3}{4}$, etc.

In this simple case the curve has a degenerate S-shape, the first point of inflection being at the end of trial 2.

Case 4. $m = 2$, $d = 3$, $k = 2$, but with our standard sampling assumption. The same language example as was used for Case 3 will work here. The only change from Case 3 is in the sampling procedure. We now sample independently to begin each block of 2 trials. Thus every odd-numbered trial has probability $\frac{1}{2}$ of being either of the two commands, and the even-numbered trials must be different from the preceding trial, i.e., must sample the other command with probability 1. The tree, from which we can derive the mean learning curve, for four trials is shown in Figure 3. What is unusual and conceptually important about this case is the *decrease* in the mean probability of a correct response on trial 4. It is a consequence of the wrong association in trial 2 of $g\,s$ not producing an incorrect response on trial 3 when the command *Get screw!* is repeated, but leading to an incorrect response on trial 4 when the command *Get nut!* is given. This

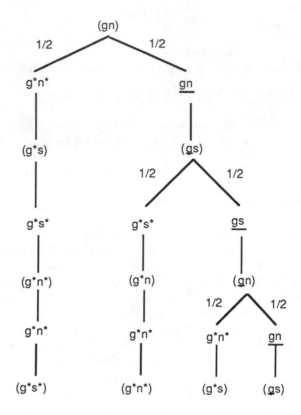

FIGURE 2. Partial Tree for Case 3. $m = k = 2$, $d = 3$.

pattern of decrease, though of smaller magnitude, occurs on every even-numbered trial. In simple homogeneous stimulus conditions such decreases in the mean learning curve never occur, but they are natural and to be expected in language learning. See Figure 4 for the mean learning curve of Case 4. Numerically, the theoretical results are $p_1 = p_2 = 0$, $p_3 = .75$, $p_4 = .625$, $p_5 = .891$, $p_6 = .859$, $p_7 = .953$, $p_8 = .941$.

Case 5. $m = 4$, $d = 2$, $k = 1$. From the parameters of this case it sounds simpler, but in fact it has, as well as two denoting words, two nondenoting words, namely *please* and *now*. Thus the four sentences of Case 5 are *Please forward!, Now stop!, Now forward!* and *Please stop!* What makes the learning more complicated in this case and is an important concept for language learning is that the number of denoting symbols in the internal representation is in every case, that is, in each of the four sentences less than the number of the words in the sentence. So there is a special opportunity for wrong associations to be formed. As might be expected the learning is slower and more complicated. The results of the first four trials are shown partially in Figure 5. Only part of the tree is drawn because it is too large

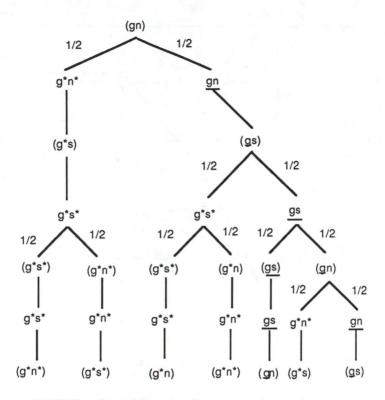

FIGURE 3. Partial Tree for Case 4. $m = k = 2$, $d = 3$.

to put on the page, but enough is given to show the nature of the results. In the tree the word that has no association to a denoting symbol at a given stage is printed at the corresponding node. Thus at the top of the figure, just below (pf), we print *please*, because *forward* has the only association. A star next to an input command like (nf) indicates a correct response with probability one. The mean learning curve is the following: $p_1 = p_2 = 0$, $p_3 = .083$, $p_4 = .3125$, $p_5 = .727$, $p_6 = .686$, $p_7 = .718$, $p_8 = .794$. This case also exemplifies the nonmonotonic mean learning curve found already in Case 4. In Section 4.3 we examine more carefully with a number of further examples the learning curves when nondenoting words are present. The examples considered there are too difficult to compute theoretically.

4.2 COMPUTATIONAL COMPLEXITY OF MEAN LEARNING CURVES

The concept of mean learning curves has been central to the study of learning in experimental psychology over many decades. In almost all experimen-

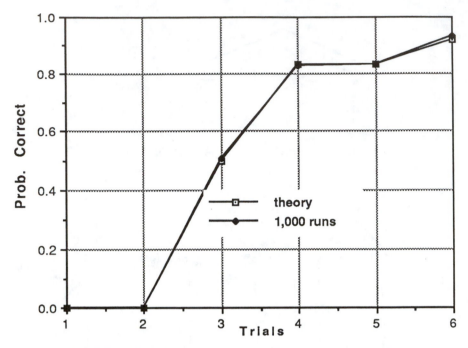

FIGURE 4. Mean Learning Curve for Case 4. $m = k = 2$, $d = 3$.

tal situations in which mean learning curves were computed theoretically, as well as presented empirically, it was quite easy in stochastic models of learning familiar from the 1950s and 1960s to compute explicitly the functional form of the mean learning curve for computation of mean learning curves in some zero-sum, two-person situations see (Suppes and Atkinson, 1960), and for the case of a continuum of responses (Suppes, 1959). Even though the conceptual framework in the two-person situations analyzed in detail in Suppes and Atkinson and the situation with a continuum of responses seem complicated, the methods developed for simpler cases generalize nicely to these more complex ones. The situation unfortunately is quite different in the case of our theory of language learning. The mean learning curves have the same conceptual meaning and force but they are computationally intractable. It is easy to see why in the examples considered in the previous section. The number of verbal commands is extremely small and from a linguistic standpoint uninteresting. On the other hand, even small samples of language lead to computationally unmanageable mean learning curves. For example, in our first robot demonstration which has Robotworld assembling a small shelf structure with nuts and bolts we trained the robot on a sample of 360 sentences in each of three languages, English, German, and Chinese, with each language being treated independently. Without intro-

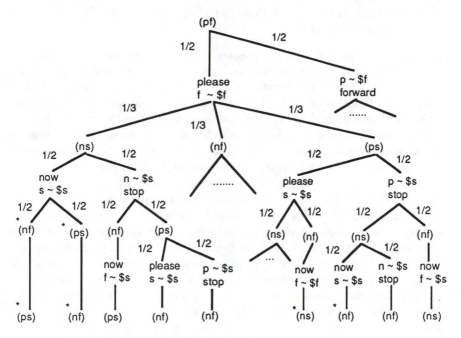

FIGURE 5. Partial Tree for Case 5. $m = 4$, $d = 2$, $k = 1$.

ducing any further special assumptions, several cycles of random sampling without replacement of the entire block of 360 trials is needed for learning with a low error rate. But as is easy to see, just from the assumptions about random sampling of the verbal commands, the number of branches of the tree structure is $m!^n$, where m is the number of distinct verbal commands and n is the number of blocks of trials, each block consisting of 360 trials. Even for our restricted experiment this is an intractable computation. We can state this result more generally in the form of a theorem.

THEOREM. *With random sampling of sentences, the computation of mean learning curves is not feasible.*

The proof of this theorem is immediate from the considerations just introduced above. Notice that the computations are far from polynomial as a function of m.

4.3 CONJECTURE: POLYNOMIAL BOUND ON MEAN LEARNING RATE

Although we are not able to theoretically compute the mean learning curves, extensive simulation studies lead us to the conjecture that in almost all cases the learning rate for a sample of m verbal commands is a polynomial function of m. We have in mind here that a rather strict criterion of learning is used, for example, at least 95 per cent correct responses.

Here are the series of examples which led us to the conjecture that in most cases the learning rate is quite fast even though the simplest theoretical concept, the mean learning curve, cannot be theoretically computed. In these examples, adding to our earlier notation of m, d and k, we now add $\bar{d} =$ the number of distinct nondenoting words and $\bar{k} =$ average number of nondenoting words per command. As in the previous cases, in the examples we consider, k and $\bar{k}$ are not simply averages but constants. We also emphasize that the concept of denoting here is a very restrictive one. It refers just to denotations within the semantic categories of actions, objects, properties and spatial relations. It is important that we have not had to introduce this concept in our learning axioms, but the presence of a large number of nondenoting words certainly has an impact on the rate of learning. This is why we have constructed examples with as many nondenoting as denoting words. We number the cases consecutively with those of Section 4.1.

Case 6. $m = 4$, $d = 4, \bar{d} = 4$, $k = 2$, and $\bar{k} = 2$. An example of this case are the following four verbal commands:

Please turn the screw!
Please get a nut!
Now turn the nut!
Now get a screw!

The mean learning curve for this case, computed numerically from the average of 100 individual sample paths, 1000 and 10,000 is shown in Figure 6. All three curves reach the criterion of 95 per cent correct in about 24 trials. Because the learning rate took more than 2^m trials to learn the four sentences we were hopeful that larger examples would confirm this rate, but the next cases show that this was mistaken.

Case 7. $d = \bar{d} = m = 6$, $k = \bar{k} = 3$, $r = 2$. Here we have also introduced a new parameter r to indicate the number of times each word is repeated in the sentences being learned. We held r to 2 in the expectation that this would slow the rate of learning. In this example and in the others that follow, we artificially constructed the language to test the rate of learning, although of course it would be possible to substitute real natural language commands for the structures we have studied. But the point here is really just to study the rate of learning for given types of structures. So we lay the language out artificially using the capital letters at the beginning of the alphabet for denoting words and the capital letters at the end of the

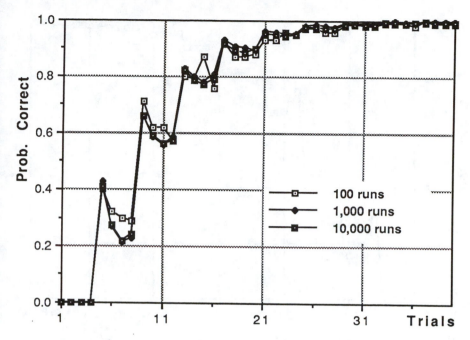

FIGURE 6. Mean Learning Curve for Case 6. $m = d = \bar{d} = 4$, $k = \bar{k} = 2$.

alphabet for nondenoting words. Here are the two sets of words:

$$d \quad = \quad |A\ B\ C\ D\ E\ F|$$
$$\bar{d} \quad = \quad |U\ V\ W\ X\ Y\ Z|$$
$$U\ C\ Y\ F!$$
$$X\ A\ V\ E!$$
$$W\ B\ V\ F!$$
$$W\ C\ Z\ E!$$
$$X\ B\ Y\ D!$$
$$U\ A\ Z\ D!$$

The learning rate for this case with 1000 individual runs as the basis for the numerical computation for the mean learning curve is shown in Figure 7. As can be seen from the figure, approximately, the 95 per cent criterion of learning was reached on average in about 42 sentences which is far below 2^6.

Case 8. $d = \bar{d} = m = 8$, $k = \bar{k} = 4$, $r = 2$. The artificial sentences were constructed in the same fashion as Case 7. The mean learning curve constructed from 1000 individual sample paths is shown in Figure 8 where the criterion of 95 per cent correct is reached at about trial 61.

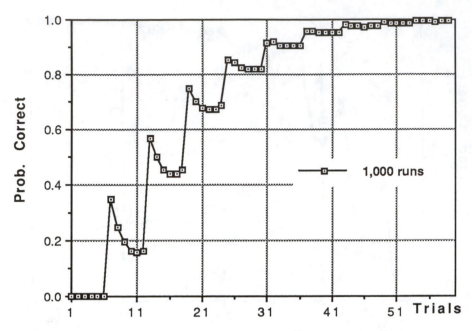

FIGURE 7. Mean Learning Curve for Case 7. $d = \bar{d} = m = 6$, $k = \bar{k} = 3$, $r = 2$.

Case 9. $d = \bar{d} = m = 6$, $k = \bar{k} = 3$, $r = 3$. In this case we increase the number of words in each sentence to 6 with each word therefore occurring in three sentences. In two of the sentences each denoting word cooccurred with the same nondenoting symbol, but not in the third occurrence. We conjectured that this positive correlation, which was of course not 1, would lead to a slowing of the learning, but it did not have a large effect as can be seen from Figure 9. The 95 per cent criterion of learning is reached at about 50.

Case 10. Finally, we describe briefly, the results for the learning of 360 Chinese verbal commands as preparatory training for executing the demo construction by Robotworld. The learning results were essentially the same for English and German. In this case the learning was accelerated by labeling the nondenoting words, which were then excluded from the probabilistic association. With this labeling, as can be seen from Figure 10, the 95 per cent learning criterion was reached in less than one cycle of the 360 verbal commands, in fact, in about 100 trials.

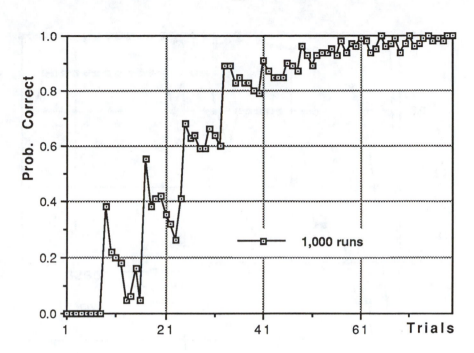

FIGURE 8. Mean Learning Curve for Case 8. $d = \bar{d} = m = 8$, $k = \bar{k} = 4$, $r = 2$.

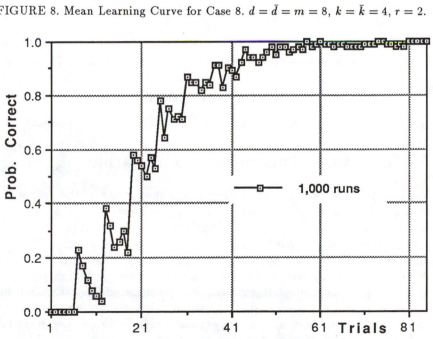

FIGURE 9. Mean Learning Curve for Case 9. $d = \bar{d} = m = 6$,
$k = \bar{k} = 3$, $r = 3$.

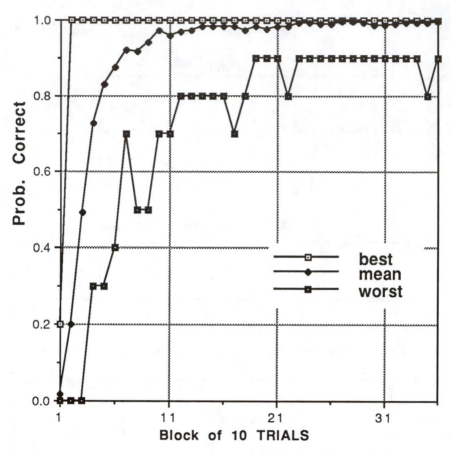

FIGURE 10. Mean Learning Curve for Case 10. 360 Chinese Commands.

5   Grammars Constructed from Learning

We have just begun the investigation of the complexity of the grammars that can be learned from a finite sample of sentences. In a sense, it is better to say grammars *constructed* from learning a fragment of a language, because the robot does not learn an explicitly formulated grammar fixed in advance, but constructs one on the basis of learning some initial fragment of a language. In any case, we present here only our first preliminary results. We stress that the context-free grammars are generated from scratch from the grammatical forms learned by the robot. They are not constructed as an extension or transformation of a grammar given *a priori*, i.e., prior to, and separate from, the learning process.

To generate the grammars we consider here, two more axioms are needed, the recursive axiom for term association mentioned earlier but not stated,

---

121

and the axiom of term form substitution.

To make the statement explicit we need to extend the earlier discussion of association of words with internal symbols to include terms, both of the given natural language and of the internal language. Formally, such associations of terms, $t \sim \tau$, were already used in the statement of Axioms 6, 7 and 8, but explicit explanation was given only for the special case of t being a single word, all that is essential for developments up to this point.

First, if σ is an internal representation of a verbal command s, the denoting *atoms* of σ are just the LISP atoms that are the internal symbols associated with words of the given natural language. By a *minimal* LISP *expression* of σ containing given occurrences of denoting atoms $\alpha_1 ..., \alpha_n$ in σ we mean the smallest LISP expression, as defined at the beginning of Section 3, containing the atoms $\alpha_1 ... \alpha_n$. For example, if $s = $ *Get the red screw!*, and $s \sim \sigma = $ *(fa1 \$g (io (fo \$r (fo \$s *))))*, then there are three occurrences of denoting atoms in σ, one of \$g, one of \$r and one of \$s. The minimal LISP expression containing the one occurrence of \$s in σ is *(fo \$s *)*.

We postulate a separate association relation for terms — separate from the association relation $\sim$, although we use the same symbol. Thus stored in memory we have

$$screw \sim (fo\ \$s\ *)$$

for the above example, as well as $screw \sim \$s$ stored under word associations. Also a string w of s is *purely nondenoting (relative to $s \sim \sigma$)* iff no word of w corresponds to each occurrence of a denoting atom in σ.

Second, given a verbal command $s \sim \sigma$ and t a nonempty substring of s, the string t is a *pure complete denoting term (relative to s and σ)* iff (i) every word of t has a corresponding association to each occurrence of a denoting atom in σ, where *corresponding* means in accordance with the association relation $s \sim \sigma$, (ii) the minimal LISP expression τ containing the corresponding occurrences of all the denoting atoms of σ associated with words of t contains no other occurrences of denoting atoms. We now use the concepts introduced to state the recursive axiom needed.

Axiom 11. (*Term Association*). First, if $a \sim \alpha$ and $\alpha \in P$ then $a \sim$ *(fo α *)*. Second, if (i) *awt* is a substring of a verbal command s, with $s \sim \sigma$ and $a \sim \alpha$, (ii) t is a pure complete denoting term of s with $t \sim \tau$, (iii) w is the substring, possibly empty, of purely nondenoting words between a and t, (iv) $\tau'(\alpha, \tau)$ is the minimal LISP expression of σ containing the occurrences of α and denoting atoms in τ corresponding to occurrences of denoting words in t and $\tau'(\alpha, \tau)$ contains no other occurrences of denoting atoms, then *awt* $\sim \tau'(\alpha, \tau)$. Moreover, the axiom applies to *twa* with w following t.

In the first part of the Axiom it follows from the category structure of the internal language that *(fo α *)* $\in O$. Notice that as the axiom is formulated in terms of pure complete denoting terms, in the intended correct associations, the atomic term *red* in the command *Get the red screw!* discussed above, is not a pure complete denoting term because the minimal LISP expression containing this occurrence of $r also contains $s.

Axiom 11 is the most complicated single association axiom we state and one that in a further development of the theory might be simplified. We did not find easy ways to do so in the present setup. On the other hand, the role of Axiom 11 is central to recursively generating from a fixed finite sample of sentences a grammar that can comprehend an infinite set of sentences.

We also need a new axiom on substituting term forms, so, for example, in the sentential form A_1 *the O!* we can substitute the form *PO* for *O*, since $PO \subseteq O$, to derive the new sentential form A_1 *the P O!*

Axiom 12. (*Term form Substitution*). If $g(X) \sim \gamma(X)$, $g'(Y_1,...,Y_n) \sim \gamma'(Y_1,...,Y_n)$ and $g'(Y_1,...,Y_n) \subseteq X$, then $g(g'(Y_1,...,Y_n)) \sim \gamma(\gamma'(Y_1,...,Y_n))$.

We can illustrate the nature of the recursion by the following example, where we show the LISP internal representations, as well as use both Axioms 11 and 12.

Assume that, on the basis of previous experience, memory holds the following associations:

$$P \quad \sim \quad (fo\ P\ *) \tag{1}$$
$$A_1\ the\ O! \quad \sim \quad (fa1\ A_1\ (io\ O)) \tag{2}$$
$$get \quad \sim \quad \$g \tag{3}$$
$$screw \quad \sim \quad \$s \tag{4}$$
$$plate \quad \sim \quad \$p \tag{5}$$
$$black \quad \sim \quad \$b \tag{6}$$
$$large \quad \sim \quad \$l \tag{7}$$
$$right \quad \sim \quad \$r \tag{8}$$

Coerce:

$$Get\ the\ black\ nut! \quad \sim \quad (fa1\ \$g\ (io\ (fo\ \$b\ (fo\ \$n\ *)))) \tag{9}$$
$$Get\ the\ nut\ which\ is\ large! \quad \sim \quad (fa1\ \$g\ (io\ (fo\ \$l\ (fo\ \$n\ *)))) \tag{10}$$
$$Move\ right\ of\ the\ nut! \quad \sim \quad (fa1\ \$m\ (fp\ \$r\ (io\ (fo\ \$n\ *)))) \tag{11}$$

Term association applied to (9) - (11) yields:

$$nut \quad \sim \quad (fo\ \$n\ *) \tag{12}$$
$$black\ nut \quad \sim \quad (fo\ \$b\ (fo\ \$n\ *)) \tag{13}$$
$$nut\ which\ is\ large \quad \sim \quad (fo\ \$l\ (fo\ \$n\ *)) \tag{14}$$
$$right\ of\ the\ nut \quad \sim \quad (fp\ \$r\ (io\ (fo\ \$n\ *))) \tag{15}$$

G.F.generalization applied to (13) - (15) yields:

$$P\ O\ \ \sim\ \ (fo\ P\ O) \tag{16}$$
$$O\ which\ is\ P\ \ \sim\ \ (fo\ P\ O) \tag{17}$$
$$R\ of\ the\ O\ \ \sim\ \ (fp\ R\ (io\ O)) \tag{18}$$

Category generalization applied to (12) - (15) yields for purposes of computation, but not permanent storage in memory:

$$nut\ \in\ O \tag{19}$$
$$black\ nut\ \in\ O \tag{20}$$
$$nut\ which\ is\ large\ \in\ O \tag{21}$$
$$right\ of\ the\ nut\ \in\ P \tag{22}$$

From (19) - (22) we get

$$P\ \subseteq O \tag{23}$$
$$P\ O \subseteq O \tag{24}$$
$$O\ which\ is\ P \subseteq O \tag{25}$$
$$R\ of\ the\ O \subseteq P \tag{26}$$

The robot can use (1) - (26) to comprehend the more complex sentence

$$Get\ the\ black\ screw\ which\ is\ right\ of\ the\ large\ plate! \tag{27}$$

The parsing and comprehension of (27) will use the small context-free grammar $\mathcal{G}$ generated by association and coercion as expressed by (1), (2), (16) - (18) and (23) - (26), which written in conventional style, but without rewrite rules for terminal words, has the following form:

$$A\ \rightarrow\ A_1\ the\ O!$$
$$O\ \rightarrow\ P$$
$$O\ \rightarrow\ P\ O$$
$$O\ \rightarrow\ O\ which\ is\ P$$
$$P\ \rightarrow\ R\ of\ the\ O$$

The first step is to identify the words of (27) that are associated with internal denoting symbols. The second step is to look up the categories of the denoting words of (27) to generate the grammatical form

$$A_1\ the\ P_1\ P_2\ which\ is\ R\ of\ the\ P_3\ P_4! \tag{28}$$

Using a standard parser for the context-free grammar $\mathcal{G}$ we are able to parse (28). We then use (1), (2) and (16) - (18) to generate the internal representation form associated with (28):

$$(fa1\ A_1\ (io\ (fo\ P_1\ (fo\ (fp\ R\ (io\ (fo\ P_3\ (fo\ P_4\ *))))(fo\ P_2*))))) \tag{29}$$

Term specification then yields at once the internal representation of (27), which the robot can then execute:

$$(fa1 \ \$g \ (io \ (fo \ \$b \ (fo \ (fp \ \$r \ (io \ (fo \ \$l \ (fo \ \$p \ *))))))(fo \ \$s \ *))))) \quad (30)$$

6 Comments on Related Work

We have found most relevant to our research the work of Jerome Feldman and his colleagues at the International Institute for Computer Science in Berkeley and the work of Jeffrey Siskind at M.I.T.

With the approach of the Feldman group (Feldman *et al.* 1990; Hollbach *et al.* 1990; Regier, 1990; Stolcke, 1990) we share the focus on spatial-relation talk. But whereas the Feldman group's work is restricted to the study of static spatial relations in a 2D environment we, in addition, consider dynamic spatial relations in a 3D environment. As a consequence, the Feldman group has no action expressions.

While the Feldman group deals with the acquisition of spatial relations, we are focussed exclusively on the learning of the words for spatial relations and assume that the notion has already been acquired by the learner.

Feldman's group emphasizes the point that their learning proceeds without negative evidence (in line with standard views on the child's learning of syntax but not actions); in contrast we use coercion to correct mistaken actions on the part of the learner. Correspondingly they use a declarative language; we use an imperative language.

The Feldman group does not presuppose any linguistic knowledge on the part of the learner and grammatical knowledge is acquired by a connectionist framework. In our approach, some linguistic knowledge is presupposed by endowing the learner with an internal language.

Siskind (1990) develops an operational system MAIMRA (extended to DAVRA in Siskind (1991a,b)) that serves the purpose of language understanding, language generation, and language acquisition at the same time. Since our system so far does not handle generation, we only focus our comparison on the acquisition and comprehension aspects of Siskind's systems.

MAIMRA constrains the possible mappings between natural language structures and scenarios (sequences of events) by three components: a parser, a linker, and an inference component.

Unsurprisingly, there are certain features shared by Siskind's systems and our system:

1. Both systems use verbal and non-verbal stimuli as input (visual scenario and description of it in MAIMRA's case, verbal stimulus and coerced action in our case).

2. Since language learning requires presence of what the talk is about there is a common focus on visually perceptible scenes and spatial relations in particular.

3. The systems have both been exposed not only to English but also to structurally more remote languages such as Japanese or Chinese and Russian (our system).

4. Both approaches make essential use of an internal language.

5. Both systems require that each word has only one meaning (Siskind's "monosemy constraint" matches our "non-ambiguity requirement").

On the other hand, the systems are different in at least the following respects:

1. Siskind's systems acquire language on the basis of assertions, our system on the basis of commands.

2. Siskind's approach appears to require a distinction between denoting words and non-denoting words such as determiners. Initially, our system had a similar condition but we succeeded in removing it.

3. In addition to monosemy but unlike Siskind we also require that a denotation can be expressed by only one word (our "non-synonymity constraint" of Axiom 3).

4. The output of Siskind's system is a lexicon (list of words with syntactic category and meaning) in the case of MAIMRA and a lexicon together with $\bar{X}$-parameters of the language to be learnt in DAVRA, whereas our output is a grammar of the language to be learned.

5. As the language of conceptual representation Siskind uses a language of predicate logic without quantifiers but with variables (to handle argument relationships and arities) whereas our internal language is a variable-free procedural language.

6. Whereas our system assumes a one-one-mapping between denoting words of the verbal command and the denotations in the semantic representation (verbal input is semantically exhaustive), Siskind here can afford a more relaxed position: the verbal input can be less informative than the non-verbal input.

7. We start from a high-level internal representation whereas Siskind works his way up from simple state-descriptions to event-descriptions by his inference component. Siskind's takes the bootstrapping objective more seriously.

8. Siskind uses the lexical categories noun, verb, and preposition. We have carefully avoided the use of syntactic categories and have used semantic categories of action, object, property, and spatial relation instead. Of course there is a close correspondence between semantic categories and syntactic categories but there are also many examples

where relations occur as adjectives as in, e.g. *the left hole*, adverbs as in, e.g. *Turn left!*, or nouns as in, e.g. *Go to the left!* and not as prepositions. We also believe that the syntactic categories are more varied from language to language than semantic categories.

We are not suggesting that the work we have just reviewed encompasses everything relevant to our own efforts, for there are many research activities worldwide the learning of natural language. Undoubtedly we and others working on machine learning of natural language will continue to benefit in many ways from the continuing research of linguists on the structure of language and of psychologists on the acquisition of language.

REFERENCES

[1] Feldman, J. A., Lakoff, G., Stolcke, A. and S. Hollbach Weber (1990) Miniature Language Acquisition: A touchstone for Cognitive Science. International Computer Science Institute, Berkeley Ca.

[2] Hollbach Weber, S. and A. Stolcke (1990) L_0: A Testbed for Miniature Language Acquisition. TR-90-010, International Computer Science Institute, Berkeley Ca.

[3] Regier, T. (1990) Learning Spatial Terms without explicit negative evidence. TR-90-057, International Computer Science Institute, Berkeley Ca.

[4] Siskind, J. M. (1990) Acquiring core meanings of words, represented as Jackendoff-style conceptual structures, from correlated streams of linguistic and non-linguistic input. *Proceedings of the 28th Annual Meeting of the Association of Computational Linguistics*, 143-156.

[5] Siskind, J. M. (1991a) Dispelling Myths about Language Bootstrapping. *AAAI Spring Symposium Workshop on Machine Learning of Natural Language and Ontology*, Stanford.

[6] Siskind, J. M. (1991b) Naive Physics, Event Perception, Lexical Semantics and Language Acquisition. *AAAI Spring Symposium Workshop on Machine Learning of Natural Language and Ontology*, Stanford.

[7] Stolcke, A. (1990) Learning Feature-based Semantics with Simple Recurrent Networks. TR-90-015, International Computer Science Institute, Berkeley Ca.

[8] Suppes, P. (1959) A linear model for a continuum of responses. In R. R. Bush & W. K. Estes (Eds.), *Studies in mathematical learning theory*. Stanford: Stanford University Press, pp. 400-414.

[9] Suppes, P. and R. C. Atkinson (1960) Markov learning models for multiperson interactions. Stanford: Stanford University Press, 296 pp.

Part II
Forecasting and Arms Race

Nonlinear Forecasting, Chaos and Statistics

M. Casdagli

Many natural and experimental time series are generated by a combination of coherent, low-dimensional dynamics and stochastic, high-dimensional dynamics. A famous example is the sunspot time series. In the first part of this paper a nonlinear forecasting algorithm is used to attempt to identify how much of the irregularity in an aperiodic time series is due to low-dimensional chaos, as opposed to high-dimensional noise. The algorithm is applied to experimentally generated time series from coupled diodes, fluid turbulence and flame dynamics, and compared to dimension calculations. Theoretical results concerning the limitations of such forecasting algorithms in the presence of noise are reviewed in the second part of the paper. These results use a combination of ideas from dynamical systems and statistics. In particular, the state space reconstruction problem is investigated using properties of likelihood functions at low noise levels.

1   Introduction

There has been much recent interest in the "modeling of complex phenomena", and there are many different approaches to the problem, as evidenced by this conference proceedings. In this brief review, the "complex phenomena" of interest are irregular time series data, of which several examples are given. Efforts to model and forecast such time series will be presented.

The nonlinear modeling and forecasting of time series data has a relatively recent history. The statistics community has constructed stochastic nonlinear models since about 1980, for a review see Tong [46]. Independently, the dynamical systems community has constructed deterministic nonlinear models since about 1987, see [13, 16, 17, 19, 22] and references therein. Rather than attempting to review this work here, I will instead review some recent results of mine which attempt to combine the stochastic and deterministic approaches. The applications to experimental data are preliminary; for more details see [15]. The theoretical results were obtained in collaboration with Eubank, Farmer and Gibson [14].

To fix ideas, consider the Wolfer sunspot time series $x_1, .., x_{287}$ of Fig. 1a, with its remarkable mixture of regularity and irregularity. Maxima occur roughly every 11 years, but the magnitudes of the maxima are irregular. In the absence of satisfactory physical models for this phenomenon, the question arises of how to best forecast the sunspots. Since sunspot activity affects conditions in the upper atmosphere, such forecasts are of great interest to those involved in satellite launching. Fig. 1b illustrates the *phase portrait* of the time series. Consider its generalization to a *state space reconstruction* of *delay vectors*

Martin Casdagli: Santa Fe Institute, 1120 Canyon Road, Santa Fe, New Mexico 87501

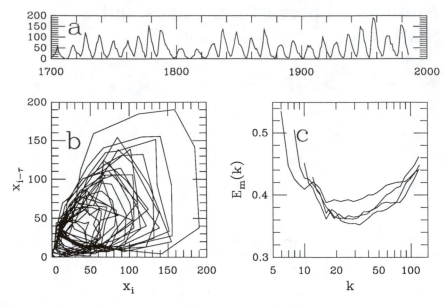

Figure 1: *Sunspot data.* (a) Annual Wolfer numbers. (b) Phase portrait with $\tau = 2$. (c) Average one-year prediction errors $E_m(k)$ for $m = 3, .., 6$ and $\tau = 1$ as a function of the neighborhood size k.

$$\underline{x}_i = x_i, x_{i-\tau}, .., x_{i-(m-1)\tau} \tag{1}$$

of *embedding dimension* m. A dynamics is induced on the reconstructed space by the time ordering of the $\underline{x}_i$, and the forecasting problem can be addressed statistically by attempting to model the dynamics to fit the observed data. This state space approach essentially reduces the forecasting problem from one of *extrapolating* the graph of Fig. 1a to the more realistic problem of *approximating* the dynamics of Fig. 1b.

In 1927, Yule developed this approach, and fitted a linear, stochastic model to the data [51]. In 1980, Tong and Lim fitted a nonlinear stochastic model to the data and obtained significantly better forecasts than existing linear models [47]. The pitfall of overfitting to the data was avoided by restricting the number of free parameters in the model using Akaike's information criterion. This resulted in a piecewise linear model consisting of two pieces on an $m = 12$ dimensional state space. Also, unlike linear models, the nonlinear model was able to capture the strong time asymmetry evident in Fig. 1a, by exhibiting a noisy limit cycle. More recently, Weigend has reported improved forecasts using a neural net model [49].

Coming from the dynamical systems community, one might argue that Fig. 1b resembles a low-dimensional strange attractor rather than a noisy limit cycle, and that nonlinear deterministic modeling might give improved forecasts, at least in the short term. Fig. 1c illustrates the results of fitting a wide range of piecewise linear models to the first half of the data, and estimating an error index $E_m(k)$ for one year ahead forecasts on the second half of the data. The *smoothing parameter* k measures the number of neighbors used in the fitting. The largest value of k corresponds to fitting a linear stochastic model, and the smallest value of k corresponds to fitting a nonlinear deterministic model. In this figure and in later figures in this paper, curves $E_m(k)$ with larger values of the embedding dimension m are computed starting at larger values of k. This fact may be used to identify the embedding dimension corresponding to each curve. Observe that the best forecasts were obtained with an intermediate value of k, yielding a nonlinear stochastic model. One might conclude from Fig. 1c that if the sunspots exhibit deterministic chaos, the chaos is of too high dimension to be approximated by fitting a fully deterministic model to the data. On the other hand, it appears that evidence for nonlinearity has been found in the data. One might conclude that although the sunspot data is of high dimension, there are coherent effects which can be profitably approximated by a low-dimensional nonlinear stochastic model.

In this paper, the forecasting algorithm used in Fig. 1c is applied to a variety of experimental time series, with the intention of identifying how much of the irregularity in the time series is due to low-dimensional nonlinear effects as opposed to high dimensional "stochastic" effects. The algorithm is intended to supplement correlation dimension calculations. It might also be useful in other applications, where optimal forecasting, noise reduction and control are of more importance. In Section 2, a more complete description of the algorithm is given, its relationship to existing work is discussed and its properties are investigated on numerical examples. In Section 3, the algorithm is applied to experimentally generated time series, and compared to dimension calculations. In Section 4, some theoretical questions raised in Section 2 are addressed. In Section 5, the main conclusions and open questions are summarized.

2 The algorithm

2.1 Description

The algorithm used to produce Fig. 1c is defined as follows; for more details see [15]. (A) Divide the time series into two separate parts; a *fitting* set $x_1, .., x_{N_f}$ and a *testing* set $x_{N_f+1}, .., x_{N_f+N_t}$. (B) Choose an embedding dimension m, a delay time τ, and a prediction time T. (C) Choose a delay vector $\underline{x}_i$ with $i \geq N_f$ for a T-step ahead prediction test. (D) Compute the distances d_{ij} of the test vector $\underline{x}_i$ from the delay vectors $\underline{x}_j$, $1 + (m-1)\tau \leq j \leq N_f - T$ in the fitting set, using the maximum norm for computational efficiency. (E) Order the distances d_{ij}, find the k nearest neighbors $\underline{x}_{j(1)}, .., \underline{x}_{j(k)}$ of $\underline{x}_i$, and fit an affine model of the form (2), where the parameters $\alpha_0, .., \alpha_m$ are fitted by ordinary least squares.

$$x_{j(l)+T} \approx \alpha_0 + \sum_{n=1}^{m} \alpha_n x_{j(l)-(n-1)\tau} \qquad l = 1, .., k \qquad (2)$$

Vary the number of nearest neighbors k at several representative values between $2(m+1)$ and $N_f - T - (m-1)\tau$. The parameters $\alpha_0, .., \alpha_m$ can be obtained by solving the normal equations for the linear system (2), which can be updated recursively as k is increased and solved using LU decomposition for computational efficiency. (F) Use the fitted model (2) to estimate a T step ahead prediction $\hat{x}_{i+T}(k)$ for the test vector $\underline{x}_i$, and compute its error $e_i(k) = |\hat{x}_{i+T}(k) - x_{i+T}|$. (G) Repeat steps (C) through (F) for all i in the testing set, and compute the normalized RMS predictor error

$$E_m(k) = (\sum_i e_i^2(k))^{1/2}/\sigma \qquad (3)$$

where σ is the standard deviation of the time series. (H) Choose the delay time τ and the prediction time T by discretion. In chaotic systems, they should not be chosen too large. In finely sampled continuous time systems, they should not be chosen too small. Vary the embedding dimension m, and study the curves $E_m(k)$ as a function of k.

2.2 Remarks

The idea of varying neighborhood sizes is intrinsic to dimension calculations, see [2, 35, 44] and references therein. Variable neighborhood sizes have been proposed for estimating embedding dimensions [3, 10, 18], and also for forecasting [7, 16, 33] (but not applied to long experimental time series as in Section 3). The properties of how prediction errors vary with the fitting size N_f have been investigated, but we only recommend this approach for data already known to be low-dimensional and relatively noise free [13, 22, 23]. The behavior of prediction errors as the prediction time T is increased has been investigated using both "direct" and "iterative" forecasting algorithms [13, 23]. The above algorithm is a "direct" algorithm, but can be modified to be iterative if desired.

Finally, it is clear that dimension calculations should be carefully interpreted when applied to nonstationary data, since stochastic non-stationary time series can appear low-dimensional to such tests; for example see [36]. For similar reasons, the forecasting algorithm above should be modified in the case of non-stationary data. In the applications to experimental data presented in Section 3, it is implicitly assumed that the experiments are sufficiently carefully controlled so that stationarity holds to a good degree of approximation. It is an open question whether improved forecasts could be obtained by modifying the forecasting algorithms to cope with the possibility of non-stationarity.

2.3 Numerical Examples

In order to test the algorithm, time series $x_1, .., x_N$ of length $N = 50000$ were generated from the Ikeda map $f : \Re^2 \to \Re^2$ defined by

$$f(x,y) = (1 + a(x \cos t - y \sin t), a(x \sin t + y \cos t)) \qquad (4)$$

where $t = 0.4 - 6.0/(1 + x^2 + y^2)$, $a = 0.9$, and measurements were generated according to the state space model (5,6), where the noise terms ϵ_i are independent identically distributed (IID) Gaussian of variance ϵ^2, and the measurement function h projects onto the first coordinate of $\underline{s}$. The attractor is of dimension about 1.7.

$$\underline{s}_{i+1} = f(\underline{s}_i) \qquad (5)$$
$$x_{i+1} = h(\underline{s}_{i+1}) + \epsilon_{i+1} \qquad (6)$$

Since this is a discrete time system, we applied the forecasting algorithm with $\tau = T = 1$. The embedding dimension was taken to be $m = 5$ and the size of the fitting and testing sets were taken to be $N_f = 40000$ and $N_t = 400$. Define the *noise level* to be the ratio of ϵ to the standard deviation of the noise free time series. Fig. 2a shows the $E_m(k)$ curves for the noise levels 0%, 2%, 20% and 100%. Dramatically improved predictions are obtained over the linear model for the noise levels 0% and 2%, and these time series are correctly identified as essentially low-dimensional. However, when the noise level is increased to 20%, the predictability is lost to the extent that one can only conclude that the time series is nonlinear. It could be either low-dimensional and noisy, or high dimensional. Finally, at a noise level of 100%, it is difficult to detect any nonlinearity at all. Fig. 2b shows the $E_m(k)$ curves for a higher dimensional example, generated from the Mackey-Glass equations (with delay parameter 30 and sampling time 6; see [13, 23]), for various embedding dimensions m. In this case there is a 3.6 dimensional attractor and a 2% noise level. Again, the dramatically improved predictions of nonlinear models over linear models allows one to correctly conclude that the data is low-dimensional.

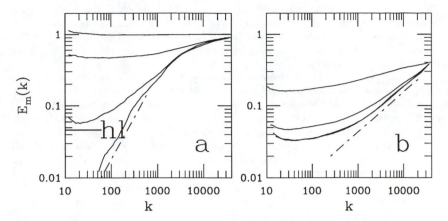

Figure 2: *Forecasting errors for numerical data.* (a) Ikeda map. $E_5(k)$ at various noise levels. The horizontal line "hl" corresponds to a *fundamental limit to predictability* due to noise amplification of the 2% noise level. (b) Mackey-Glass equation. $E_m(k)$ for $m = 4, .., 7$.

2.4 Theoretical issues

In addition to the practical utility of the above algorithm as a means of identifying low-dimensional behavior, it also has some interesting theoretical properties as follows. Firstly, suppose that the noise is negligible. Then forecasting errors are dominated by the inaccuracies in fitting a piecewise linear model to the underlying deterministic dynamics and we anticipate the scaling law

$$E_m(k) \sim C(k/N_f)^{2/D} \quad k/N_f \to 0 \tag{7}$$

where C denotes an averaged curvature of the dynamics and D denotes the dimension of the attractor; compare [13, 22]. It is typically sufficient to take an embedding dimension $m > 2D$ for (7) to hold; if $D < m < 2D$ occasional bad predictions might occur which can affect the scaling law [39]. The dashed lines in Fig. 2 were obtained by substituting the known values for D into the scaling law, and are well obeyed at low noise levels. If the attractor is multifractal, it can be argued that the precise value of D is given by D_q, where q is the solution to $(q-1)D_q = -4$, and D_q are the Renyi dimensions. The arguments are similar to those in Badii et al. [4]. If the RMS average (3) is replaced by the geometric mean, then the appropriate value of D in (7) is the information dimension D_1.

Secondly, suppose that forecasting errors are dominated by noise, rather than curvature in the dynamics. Then we expect

$$\sigma^2 E_m^2(k) \approx \epsilon^2 \langle \alpha_\epsilon^2(T, i) \rangle_i (1 + (m+1)/k) \quad N_f \to \infty \tag{8}$$

where the *noise amplification* $\alpha_\epsilon(T, i)$ is defined by the *conditional variance* of the stochastic process $X_0, X_1, .., X_N$ underlying the data according to

$$\alpha_\epsilon^2(T, i) = \mathrm{Var}(X_{i+T} | \underline{X}_i = \underline{x}_i)/\epsilon^2 \tag{9}$$

The term $(m+1)/k$ in (8) accounts for statistical problems in estimating the $m+1$ free parameters in fitting to a neighborhood of size k, and may be derived from linear time series analysis. The horizontal line "hl" in Fig. 2a was obtained by taking $\epsilon = 0.02\sigma$, ignoring the $(m+1)/k$ term, and using an approximation for the noise amplification α_ϵ^2 which we derive in Section 4.

The horizontal line "hl" represents a *fundamental limit to predictability* for *any* function approximation scheme due to noise amplification of the 2% noise level. This is because optimal predictions (in the least squares sense) for a stochastic process are obtained by predicting at the conditional mean, which results in prediction errors of variance given by the conditional variance used in (9), for example see [38]. Any function approximation scheme which attempts to approximate the conditional mean inevitably results in additional errors due to problems of statistical estimation. In this example, close to optimal forecasts have been obtained because the dynamics is relatively simple, and a large number of data points were used. We will review theoretical properties of noise amplification in Section 4.

3 Experimental data

3.1 Coupled diodes

Electrical circuits containing diodes are a useful source of experimental data for testing ideas in nonlinear dynamics [29, 34]. The forecasting algorithm of Section 2 was applied to data generated by circuits containing n coupled diodes, for $n = 2, 3, .., 8$. The complexity of the data increases with n. Since the data was sampled at the frequency of the periodic driving current, it is natural to fix the delay time $\tau = 1$ and the prediction time $T = 1$. Figs. 3a and 3b show phase portraits for the cases $n = 2$ and $n = 6$ respectively. Figs. 3c and 3d show the corresponding $E_m(k)$ curves for $m = 3, 4, .., 7$. The size of the fitting and testing sets were taken to be $N_f = 20000$ and $N_t = 400$.

Since the attractor of Fig. 3a appears multifractal, the $E_m(k)$ curves in this case were also computed using a geometric mean in place of (3); these are the dashed curves in Fig. 3c. This was the only case where we used the geometric mean. It often has poor convergence properties as the size of the testing set is increased. Deterministic (low k) models clearly give dramatic improvements in predictive accuracy over linear models, and this time series is identified as low-dimensional chaos, with an information dimension of about 3.5. By contrast, in Fig. 3d, linear models are superior to deterministic models, and nonlinear stochastic models at intermediate values of k give an average 10% improvement in predictive accuracy over linear models. In this case the time series is identified as high dimensional, but with enough coherent behavior for low-dimensional nonlinear stochastic models to give improved forecasts.

Finally, Figs. 3e and 3f show the results of dimension calculations for the two cases. The dimension estimates $\nu_m(r)$ were obtained from standard correlation sums $C_m(r)$ according to

$$\nu_m(r) = (\log C_m(r') - \log C_m(r))/(\log r' - \log r) \tag{10}$$

where r' is taken so that $C_m(r')/C_m(r) \approx 5$, following a recommendation of Theiler [45]. The embedding dimension m was varied between 2 and 10. Fig. 3e indicates low-dimensional behavior of correlation dimension about 2.5. The interpretation of Fig. 3f is more open to debate, but perhaps indicates a correlation dimension of about 4.5. We believe that in both of these cases, the forecasting algorithm has provided useful extra information to support, or call into doubt, the results of the dimension calculations.

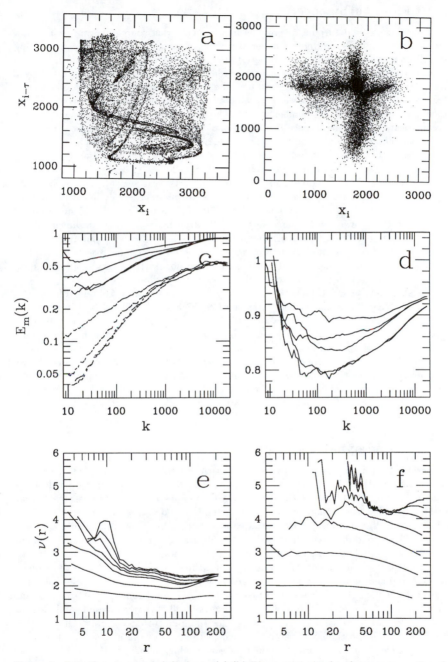

Figure 3: *Data from n coupled diodes.* (a),(b) Phase portraits for the cases $n = 2$ and $n = 6$ with $\tau = 1$. (c),(d) Average error curves for a prediction time $T = 1$. The dashed lines correspond to a geometric mean error. (e),(f) Dimension calculations with $\tau = 1$.

3.2   Fluid turbulence

Experiments in fluid turbulence provide an extremely rich source of time series data, due to the potentially enormous number of degrees of freedom available. The forecasting algorithm was applied to data sets believed to be at the two extremes of weak turbulence and strong turbulence. The first data set, illustrated in Fig. 4a, was generated by a Rayleigh Benard convection experiment near the transition to turbulence [30]. The second data set, illustrated in Fig. 4b, was generated by a Taylor Couette experiment at a highly turbulent regime (Reynolds number $\approx 10^5$). The application of the forecasting algorithm is complicated due to the absence of a natural choice for τ and T. A systematic variation of these parameters remains to be done. The choices shown in Fig. 4 seem to be reasonable. The size of the fitting and testing sets were taken to be $N_f = 20000$ and $N_t = 400$. Figs. 4c and 4d show the $E_m(k)$ curves for $m = 4, 8, .., 20$.

For the first data set, deterministic (low k) models clearly give dramatic improvements in predictive accuracy over linear models, and this time series is identified as low-dimensional chaos, with a dimension of about 3.0. This is consistent with the results of Farmer and Sidorowich on the same data set [22]. By contrast, in Fig. 4d, linear models are superior to deterministic models, and nonlinear stochastic models at intermediate values of k give only an average 3% improvement in predictive accuracy over linear models. In this case the time series is identified as high dimensional with a very small amount of coherent behavior which can be exploited by nonlinear stochastic modeling.

Finally, Figs. 4e and 4f show the results of dimension calculations obtained from (10). The embedding dimension m was varied between 2 and 10. The delay times τ were the same as those used for Figs. 4a and 4b. Also, no distances were computed between delay vectors $\underline{x}_i$ and $\underline{x}_{i'}$ with $|i - i'| < \tau$, to avoid spurious low-dimensional effects [43]. Fig. 4e indicates low-dimensional behavior of correlation dimension about 3.0. Fig. 4f indicates that the time series is probably indistinguishable from random noise. Again, the forecasting algorithm has provided useful extra information to support the results of these dimension calculations. In the case of the second data set, a small amount of extra structure was picked up by the forecasting algorithm which was not apparent in the dimension calculation.

3.3   Flame dynamics

Experiments in flame dynamics also provide a rich source of time series data. They have the advantage that the spatio-temporal motion can be readily captured on videotape. Flame dynamics data have been analyzed using techniques of spectral analysis by Gorman et al. [28]. The forecasting algorithm was applied to two data sets. The first data set, illustrated in Fig. 5a, was generated near what is believed to be a Silnikov bifurcation to chaos. The second data set, illustrated in Fig. 5b, was generated at a parameter setting giving rise to "high dimensional" unsteady cellular flames.

Again, the application of the forecasting algorithm is complicated due to the absence of a natural choice for τ and T. The choices shown in Fig. 5 seem to be reasonable. The size of the fitting and testing sets were taken to be $N_f = 15000$ and $N_t = 400$. Figs. 5c and 5d show the $E_m(k)$ curves for $m = 4, 8, .., 20$. Figs. 5e and 5f show the results of dimension calculations, with the same conventions used for the fluid turbulence data. Observe that Figs. 5d and 5f are similar to those obtained in the case of the highly turbulent fluid. Again, we conclude that in this case the time series is high dimensional with a very small amount of coherent behavior which can be exploited by nonlinear stochastic modeling.

Figs. 5c and 5e are more difficult to interpret. Nonlinear stochastic models at intermediate values of k give more than 100% improvement in predictive accuracy over linear models. The dimension calculations have not converged to a low dimension, but when

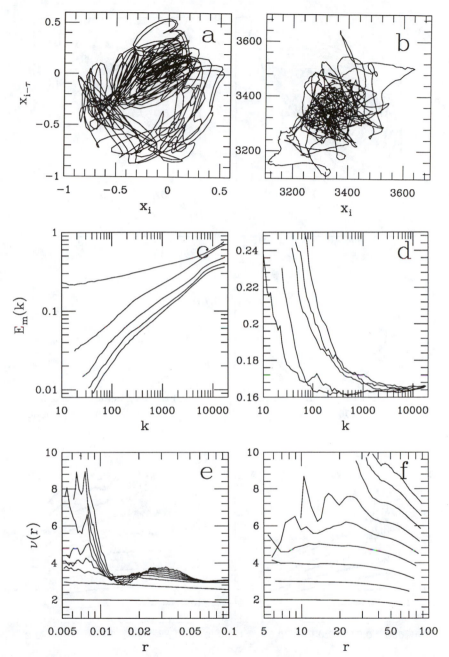

Figure 4: *Data from fluid turbulence.* (a),(b) Phase portraits for the cases of weak (resp. strong) turbulence, with $\tau = 20$ (resp. $\tau = 30$). (c),(d) Average error curves for a prediction time $T = 20$ (resp. $T = 10$) and delay time $\tau = 5$ (resp. $\tau = 2$). (e),(f) Dimension calculations with $\tau = 20$ (resp. $\tau = 30$).

140

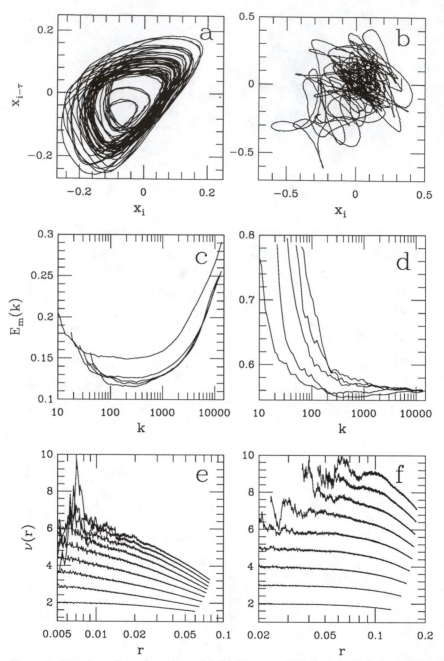

Figure 5: *Data from flame dynamics.* (a),(b) Phase portraits for what is believed to be a low (resp. high) dimensional system, with $\tau = 10$ (resp. $\tau = 100$). (c),(d) Average error curves for a prediction time $T = 10$ (resp. $T = 50$) and delay time $\tau = 2$ (resp. $\tau = 10$). (e),(f) Dimension calculations with $\tau = 10$ (resp. $\tau = 100$).

compared to Figs. 4f and 5f do not appear to be consistent with high dimensional behavior either. We have observed similar behavior of the $E_m(k)$ curves in the case of the Ikeda map example at a 20% noise level, and in n-coupled diodes where $n = 3$ or 4. At this stage we can only conclude that this time series is either low-dimensional with a moderate amount of noise (this could be dynamical noise rather than observational noise, at a level of say 20%), or of moderate dimension (say $D \approx 6$). We anticipate that this conclusion will be clarified by the study of other examples of flame dynamics, together with further experience gained on the behavior of forecasting errors and dimension calculations.

3.4   Other examples

We have applied the forecasting algorithm to other examples which will be reported in detail elsewhere [15]. For measles epidemic data, an improvement of about 15% in fore-castability over linear models was obtained using nonlinear stochastic models, confirming results of Sugihara and May [41]. So far, on the electroencephalogram, economic and meteorological data that we have analyzed, we have been unable to detect substantial nonlinear forecastability using the algorithm of Section 2. In fact, the phase portraits and error curves look even less structured than the high dimensional examples of Figs. 4b,d and Figs. 5b,d. As a general rule, when analyzing naturally occurring time series data, we have found that a less "structured" phase portrait leads to less "clear" evidence for non-linearity in forecasting algorithms. This is in contrast to dimension calculations, which, unless carefully interpreted, can lead to strong claims for low-dimensionality even in very "unstructured" phase portraits.

However, in the case of economic data, it has been found by others that if attention is focused on special segments of the time series, then there is a statistically significant amount of nonlinear forecastability [33, 50]. The improvement over linear modeling was found to be very small, and significance levels were carefully established using bootstrap-ping techniques [21], as have recently been applied to dimension calculations [8]. See also [37] for similar ideas. Nonlinear forecastability has also been detected in speech data [48], and ice age data [32].

4   Theoretical results

In this section we present a low noise limit formula for the noise amplification (9), and review its properties. This result is of importance in establishing the limitations to pre-dictability in the presence of low level noise illustrated in Fig. 2a, and applies independently of the function approximation technique used to fit the dynamics. This result can also be used to address theoretical issues in state space reconstruction. For more details see [14].

4.1   Likelihood functions

In order to develop a theory to compute noise amplification, we assume that the functional forms of f and h in (5,6) and (6) are known. Admittedly this is rarely the case in applications. Consider the problem of how to estimate the "hidden variable" $\underline{s}_i$, given f, h and the noisy delay vector $\underline{x}_i$ of dimension $m = 1 + m^+ + m^-$.

$$\underline{x}_i = (x_{i+m+\tau}, ..x_{i+\tau}, x_i, x_{i-\tau}, .., x_{i-m-\tau})^\dagger \tag{11}$$

Here "$\dagger$" denotes "transpose", and in this section we adopt the convention that states are represented by column vectors. We are usually interested in the case $m^+ = 0$, but also consider the general case since it is relevant to noise reduction. Breeden et al. have developed heuristic algorithms for solving such problems [6]. Geweke has investigated likelihood functions for the case of the tent map [27]. Following Geweke, we compute a likelihood function in the case of Gaussian IID observational noise of variance ϵ^2. Ignoring constants of proportionality, it is straightforward to show that the likelihood function $p(\underline{x}_i|\underline{s}_i)$ is given by

$$p(\underline{x}_i|\underline{s}_i) = \exp[-\parallel \underline{x}_i - \Phi(\underline{s}_i) \parallel^2 /2\epsilon^2] \qquad (12)$$

where "Takens embedding map" Φ (see [42]) is defined by

$$\Phi(\underline{s}) = (hf^{m^+\tau}(\underline{s}), .., h(\underline{s}), .., hf^{-m^-\tau}(\underline{s}))^\dagger \qquad (13)$$

The method of maximum likelihood dictates that the unknown state $\underline{s}_i$ should be estimated so as to give the maximum likelihood $p(\underline{x}_i|\underline{s}_i)$ of observing the delay vector $\underline{x}_i$. The estimation problem is thus solved by minimizing the "sum of squares" quantity $\parallel \underline{x}_i - \Phi(\underline{s}_i) \parallel^2$ with respect to $\underline{s}_i$. Figs. 6a-d illustrate the structure of this problem as m^- and m^+ are varied for the Ikeda example of Section 2, with noise level 2%. The white cross denotes the true state of the system. The darker regions correspond to smaller values of the "sum of squares" quantity $\parallel \underline{x}_i - \Phi(\underline{s}_i) \parallel^2$. Observe that for large values of m, local minima become extremely numerous. Also, as m^- (resp m^+) is increased it is possible to infer the value of $\underline{s}_i$ in the stable (resp. unstable) direction. Various strategies for finding the global minimum in such problems have been considered in [6, 24, 27].

In [14] we have concentrated on how sharp the likelihood function is at the global minimum, once it has been found. Figs. 6e and 6f show likelihood functions for noise levels 2% and 100% respectively. In Fig. 6e the horizontal axes have been blown up by a factor of about 70 relative to Fig. 6f. Observe that for the low noise level, the likelihood function is very simple, but is much more complicated at the high noise level. It can be argued from extensions to Takens' theorem [39] that, for sufficiently low noise levels ϵ and $m \geq 3$ and almost any $\underline{x}_i$, only one of the local minima contributes significantly to the likelihood function for this example. Moreover, in the limit as $\epsilon \to 0$, it can be shown that the likelihood function tends to a Gaussian with covariance matrix $\epsilon^2 \Delta(i)$, where the *distortion matrix* $\Delta(i)$ is given by

$$\Delta(i) = (D\Phi^\dagger D\Phi)^{-1} \qquad (14)$$

and $D\Phi$ is evaluated at $\underline{s}_i$. By transforming the Gaussian by the linear map $DhDf^T$, and convolving with observational noise, we derive the following result for the noise amplification

$$\lim_{\epsilon \to 0} \alpha_\epsilon^2(T,i) = 1 + DhDf^T\Delta(i)(Df^T)^\dagger Dh^\dagger \qquad (15)$$

We used this formula to obtain the horizontal line "hl" in Fig. 2a.

4.2 Properties of distortion

The noise amplification depends on the reconstruction parameters m and τ solely through the distortion matrix Δ. In order to investigate state space reconstruction problems, we therefore focus attention on Δ, since it is independent of T. To investigate the properties of the distortion matrix, we have found it convenient to condense the information into one real number, the *distortion* δ, defined by

$$\delta(i) = \sqrt{\mathrm{trace}\Delta(i)} \qquad (16)$$

The term "distortion" was originally used for a closely related quantity defined by Fraser [25], and insight into the state space reconstruction problem was gained by estimating the distortion from simulated time series data. By contrast, we are able to conveniently

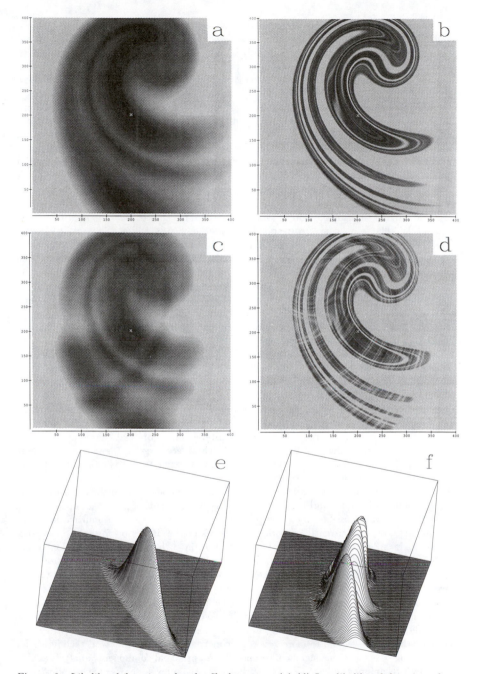

Figure 6: *Likelihood functions for the Ikeda map.* (a)-(d) Log-likelihood functions for the cases $(m^+, m^-) = (0, 2)$, $(m^+, m^-) = (0, 6)$, $(m^+, m^-) = (2, 2)$ and $(m^+, m^-) = (6, 6)$ respectively, as a function of the two dimensional hidden variable $\underline{s}_i$. (e),(f) Likelihood functions for the case $(m^+, m^-) = (0, 2)$, for noise levels 2% and 100% respectively.

compute the distortion directly from the state space model (5,6), using equations (14) and (16). It should also be noted that the distortion matrix is closely related to a statistical quantity called "Fisher's information matrix" and a control theoretic quantity called an "observability matrix". The following discussion may be viewed as summarizing our results for such matrices for chaotic systems with scalar measurement functions and low levels of observational noise.

Figs. 7a-c summarize the results of computing the distortion for the Lorenz equations, with measurement function $h(x, y, z) = x$, and sampling rate $\tau_s = 0.01$. Fig. 7a illustrates how $\delta(i)$ depends on i for the case $\tau = 2\tau_s$ and $m = 5$. Whenever x_i is close to zero, the distortion becomes very large. This phenomenon has an intuitive interpretation: in this example, if x is close to zero, the coupling to the z dynamics becomes small, and in the presence of noise, very little can be inferred about the value of z. The distortion is able to quantify this intuition concerning *information flow* between variables. In Figs 7b, and 7c the value of i is fixed. Fig 7b illustrates how δ depends on τ (measured in the natural units of time for the Lorenz system) for the three choices $(m^+, m^-) = (0, 2), (0, 4)$ and $(4, 4)$. The lower curves correspond to higher embedding dimensions m. Evidently small values of τ should be avoided; they are said to provide *redundant* information due to over-sampling. At larger values of τ the behavior is complicated; this has also been observed in information theoretic criteria for choosing optimal values of τ in dimension calculations [26]. Fig. 7c illustrates how δ depends on the embedding dimension m, for two cases with $\tau = 0.01$. The dashed curve is for the case $m^+ = m^- = (m - 1)/2$, and the asymptotic behavior, $\delta \sim m^{-1/2}$, is consistent with the law of large numbers. In the $m^+ = 0$ case, a plateau is reached; further increasing the value of m is said to provide *irrelevant* information, due to chaotic behavior.

Fig. 7d shows how the limiting distortion $\lim_{m \to \infty} \delta_m$ depends on the Liapunov exponent λ and the dimension D for an exactly solvable example. In this example, the dynamical system consists of decoupled piecewise linear maps, and the measurement function is defined by averaging the outputs of the maps; for details see [14]. Observe the explosion in the distortion as λ and D increase. We conjecture that this is the result of the impossibility of making D measurements which avoid both the problems of redundancy and irrelevancy mentioned above. An interesting theoretical consequence is that a scalar time series generated from a system with a very large distortion behaves more like a *random process* than deterministic chaos, in the sense that it is unpredictable over times much shorter than the Liapunov time.

We have developed scaling laws which describe the dependence of the distortion δ on τ, m, D and λ illustrated in Fig. 7, and which are applicable to a wide class of dynamical systems [14]. It is possible to modify these scaling laws in the case of multivariate time series [31]. We are currently exploring properties of the distortion for spatio-temporal systems, in order to quantify the information flow in such systems. In addition to obtaining theoretical insights into limitations to predictability in the presence of noise, we have also developed a theory of optimal state space reconstruction, which appears to have some practical implications. Observe that the plateau in Fig. 7c occurs at an embedding dimension $m \approx 150$, which is a high value to use for practical state space reconstruction. The question arises as to whether it is possible to compress most of this information into a smaller number of variables, using a transformation $\Psi : \Re^m \to \Re^{m'}$ with $m' \ll m$ which takes delay vectors to more general reconstructed states. Practical techniques for attempting to achieve this include filtering [40] and principal components analysis (PCA) [9]. Both techniques amount to a linear choice for the map Ψ. Fraser has demonstrated on an example that nonlinear choices for Ψ can be highly advantageous [25].

A theory for how to make an optimal choice for Ψ can be developed by showing that

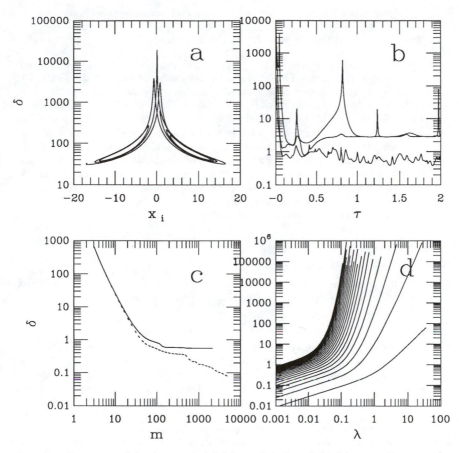

Figure 7: *Properties of the distortion* (a)-(c) as a function of the delay coordinate x_i, the time delay τ and the embedding dimension m for the Lorenz equations. (d) as a function of the Liapunov exponent λ for a solvable example of dimension $D = 2, 4, .., 40$. Higher curves correspond to larger values of D.

(14) can be generalized from delay coordinates to general reconstructions Ψ according to

$$\Delta(i) = (D\Phi^\dagger D\Psi^\dagger (D\Psi D\Psi^\dagger)^{-1} D\Psi D\Phi)^{-1}. \tag{17}$$

It can be shown that the eigenvalues of Δ are simultaneously minimized by choosing $D\Psi(t) = U^\dagger$, where U is obtained from the SVD of $D\Phi$ evaluated at $\underline{s}_i$, i.e. $D\Phi = UWV^\dagger$. Since $\underline{s}_i$ varies, this means that Ψ is a globally nonlinear reconstruction. In fact it is possible to estimate $U^\dagger$ directly from a time series, without knowing an explicit form for the map Φ. We refer to this nonlinear reconstruction technique as *local PCA*. It has already been used in modeling by others for different reasons [11, 48]. The above analysis provides a solid theoretical reason for using the technique, although we have admittedly ignored numerous complications such as the effects of estimation error, correlated noise and dynamical noise. In the case of the Lorenz system, there are theoretical reasons for believing that linear reconstruction techniques are close to optimal, and indeed we found that local PCA did not provide an advantage over standard, global PCA. It would be interesting to know if similar results hold for Fraser's example [25].

4.3 Extensions

The assumption of IID observational noise used above is clearly unrealistic for finely sampled continuous time systems. However the expression (14) for the distortion matrix can be generalized to the case of correlated Gaussian noise according to

$$\Delta = (D\Phi^\dagger \Sigma^{-1} D\Phi)^{-1} \tag{18}$$

where the observational noise terms ϵ_i are assumed to be correlated according to $\langle \epsilon_i \epsilon_j \rangle = \Sigma_{ij}$. We refer to noise added to the underlying dynamics (i.e. the right hand side of (5)) as *dynamic noise*, and formulae for the distortion matrix can also be derived in this case. In general, the expressions are considerably more involved, and preliminary results indicate that they may have some different properties compared to the observational noise case [12].

It is straightforward to solve a simple example which illustrates some of the differences between observational noise and dynamic noise. Consider the problem of estimating the parameter μ in the logistic map (19) from observations $x_0, x_1, .., x_{m-1}$, where ϵ_i is IID Gaussian dynamic noise.

$$x_{i+1} = 1 - \mu x_i^2 + \epsilon_i \tag{19}$$

Then it is straightforward to show that the likelihood function in the case of dynamic noise is given by

$$p(x_0, x_1, .., x_{m-1}|\mu) = \exp -\frac{1}{2\epsilon^2} \sum_{i=0}^{m-2} [x_{i+1} - (1 - \mu x_i^2)]^2 \tag{20}$$

The exponent is quadratic in μ, and is readily minimized.

By contrast, the expression for the likelihood function for this problem in the case of observational noise contains iterates of the logistic map, and is more complicated. A log-likelihood function is illustrated in Fig. 8. It was obtained by taking $\underline{s} = (x, \mu)$, $f(x, \mu) = (1 - \mu x^2, \mu)$ and $h(x, \mu) = x$ in (12) and (13). Note that it is natural to include the initial value of x_0 as an unknown parameter in the case of observational noise; see also Geweke [27]. It would be interesting to know if the sharpness of the maximum is more well defined for the case of observational noise. This may underlie observations that it is advantageous to fit models to iterates of the dynamics as well as to one-step dynamics, when determining free parameters [1, 6].

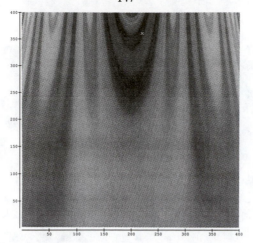

Figure 8: *A Log-likelihood function for the logistic map* as a function of (x_0, μ) for a noise level of 2%. The white cross denotes the true state of the system. The darker regions correspond to smaller values of the log-likelihood function in the case $(m^+, m^-) = (7, 0)$.

Finally, there has been recent interest in modeling and forecasting in the case of "symbolic noise" [5, 20, 52]. Symbolic noise arises from the finite resolution of measuring instruments. In theoretical studies, this process is often modeled by recording symbols, typically 0 or 1, depending on the location of the state of the system at time i. We have initiated investigations of this problem from the point of view of likelihood functions for some examples. The results are illustrated in Fig. 9.

Fig. 9a is for the case of the logistic map (19), and should be compared to Fig. 8. In this example the symbols 0 or 1 are recorded depending on whether $x_i \leq 0$ or $x_i > 0$. A sequence of $m = 8$ symbols was recorded from iterations of (x_0, μ_0), corresponding to the white cross. Each point from a 400×400 grid was tested for the symbol sequence that it generates, and shaded according to the number of correct symbols. The more symbols that are correct, the darker the shading is. The 8 measurements restrict the unknown initial condition to the small black region. The structure of Fig. 9a has a natural interpretation in terms of the symbolic dynamics of the logistic map, and goes some way towards explaining the global structure of Fig. 8 in the Gaussian observational noise case.

Fig. 9b is for the case of the toral automorphism

$$(x_{i+1}, y_{i+1}) = (ax_i + by_i, cx_i + dy_i) \mod (1, 1) \qquad (21)$$

with $a = b = c = 1$ and $d = 0$. In this example the symbols 0 or 1 are recorded depending on whether $x_i \leq 1/2$ or $x_i > 1/2$, and the problem is to identify the initial state (x_0, y_0) from a symbol sequence with $(m^+, m^-) = (2, 2)$. The solution is illustrated in Fig. 9b. By contrast to the case of the logistic map, this example is hyperbolic. The two symbols induce a Markov partition, and the initial state can be inferred with exponential accuracy as m^+ and m^- are increased. Fig. 9c is for the case of a toral automorphism with $a = 2$ and $b = c = d = 1$, known as "Arnold's cat map". In this case, the resolution of the two symbols is insufficient to infer that the initial state lies in a small connected region; the two symbols do not induce a good Markov partition. Fig. 9d illustrates that this difficulty can be avoided by increasing the number of symbols so that 0,1,2 or 3 are recorded depending on whether $x_i \in [0, 1/4), [1/4, 1/2), [1/2, 3/4)$ or $[3/4, 1)$.

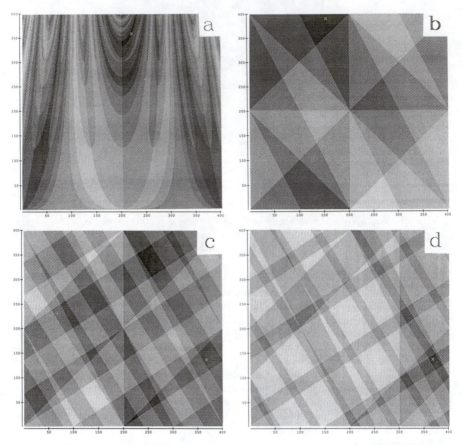

Figure 9: *Likelihood functions for symbolic noise.* (a) Logistic map. The likelihood function is uniformly concentrated on the black region of the (x, μ) plane. The white cross denotes the true state of the system. (b) Toral automorphism. (c) Cat map with two symbol recordings. (d) Cat map with four symbol recordings.

5 Conclusions

A forecasting algorithm has been presented for the analysis of experimental data which is intended to aid in the identification of low-dimensional chaos. Compared to dimension calculations, the forecasting algorithm is more CPU intensive, and contains essentially one extra free parameter: the prediction time. However, on existing workstations, we have been able to obtain encouraging results which are a useful supplement to, and often less ambiguous than, results produced by dimension calculations. We have not been misled into claiming evidence for low-dimensional chaos on wide ranges of time series data. On the other hand, evidence for low levels of nonlinearity have been identified by ourselves and other researchers using forecasting algorithms on several data sets. Some of these results may have commercial implications, and have stimulated investigations into the optimality of various state space reconstruction and function approximation techniques for forecasting. A time series competition is to be held by the Santa Fe Institute in the fall of 1991 to compare various forecasting techniques on several data sets. At this stage of research the utility of such results for the scientific enterprise is unclear. However, in cases where high levels of nonlinearity have been identified by these "statistical" techniques, these results might aid in the development or improvement of more traditional "first principles" models. In a sense, forecasting algorithms are a "solution" looking for the right problem.

A theoretical analysis of the forecasting problem in the presence of low levels of noise was also reviewed. Questions concerning information flow, state space reconstruction and limitations to predictability were addressed. The focus of attention was mostly on observational noise, and a limited range of examples were studied. It would be interesting to study a wider range of dynamical systems with a wider range of noise types using these ideas. This theoretical approach is directly applicable to the analysis of experimental time series data for which there is a known or conjectured underlying first principles model. In some experimental systems (for example coupled diodes) this may be case, and it would be interesting to try these ideas out on "realistic" models for experimental data, rather then concentrating solely on "toy" models such as the Ikeda map or Lorenz equations and numerically generated data.

I would like to thank Paul Linsay for supplying the diode data, Dan Lathrop, J.Fineberg and H.Swinney for supplying the Taylor-Couette data, and Mike Gorman, M. el-Hamdi and Kay Robbins for supplying the flame data. I would also like to thank James Theiler for supplying me with the dimension calculation code. I am grateful to partial support from the National Institute of Mental Health under grant 1-R01-MH47184-01.

References

[1] H.D.I. Abarbanel, R. Brown, and J.B. Kadtke. Prediction and system identification in chaotic nonlinear systems: Time series with broadband spectra. *Phys. Lett. A*, 18:401-408, 1989.

[2] N. B. Abraham, A. M. Albano, A. Passamante, and P. E. Rapp, editors. *Measures of Complexity and Chaos*, volume 208 of *NATO Advanced Science Institute Series*. Plenum, New York, 1989.

[3] Z. Aleksic. Estimating the embedding dimension. Technical report, Visual Sciences Centre, Australian National University, Canberra, ACT2601, 1990.

[4] R. Badii and A.Politi. Statistical description of chaotic attractors: The dimension function. *J. Stat. Phys.*, 40:725, 1985.

[5] R. Badii. Unfolding complexity in nonlinear dynamical systems. In: N. B. Abraham, A. M. Albano, A. Passamante, and P. E. Rapp, editors. *Measures of Complexity and Chaos*, volume 208 of *NATO Advanced Science Institute Series*. Plenum, New York, 1989.

[6] J.L.Breeden and A.Hubler, Reconstructing Equations of Motion from experimental data with hidden variables. *Phys. Rev. A*, 42:5817-5826, 1990.

[7] K.Briggs. Improved methods for the analysis of chaotic time series. Mathematics Research Paper 90-2, La Trobe Univ. Melbourne, Australia, 1990.

[8] W.Brock, W.Dechert and J.Scheinkman. A test for Independence based on the correlation dimension. University of Wisconsin-Madison SSRI Paper 8702, 1987.

[9] D.S. Broomhead and G.P. King. Extracting qualitative dynamics from experimental data. *Physica*, 20D:217, 1986.

[10] D.S. Broomhead, R.Jones and G.P. King. Topological dimension and local coordinates from time series data. *J.Phys. A: Math. Gen*, 20:L563-L569, 1987.

[11] D.S. Broomhead, R.Indik, A.C.Newell and D.A.Rand. Local adaptive Galerkin bases for large dimensional dynamical systems. *Submitted to Nonlinearity*, 1990.

[12] D.S. Broomhead and M.Casdagli. Unpublished research.

[13] M. Casdagli. Nonlinear prediction of chaotic time series. *Physica* , 35D:335, 1989.

[14] M. Casdagli, S. Eubank, J.D. Farmer, and J. Gibson. State space reconstruction in the presence of noise. To appear in Physica D, 1991.

[15] M. Casdagli. Chaos and deterministic versus stochastic nonlinear modeling. Preprint 1991.

[16] M. Casdagli, D. DesJardins, S.Eubank, J.D.Farmer, J.Gibson, N.Hunter and J.Theiler. Nonlinear modeling of chaotic time series: theory and applications. In J.Kim and J.Stringer, editors, *EPRI workshop on Applications of chaos* 1991 to appear.

[17] M. Casdagli and S. Eubank, editors, *Nonlinear Prediction and Modeling*. Addison-Wesley, 1991. to appear.

[18] A. Cenys and K. Pyragas. Estimation of the number of degrees of freedom from chaotic time series. *Phys. Lett. A*, 129:227, 1988

[19] J.P. Crutchfield and B.S. McNamara. Equations of motion from a data series. *Complex Systems*, 1:417–452, 1987.

[20] J.P. Crutchfield and K. Young. Inferring statistical complexity. *Phys. Rev. Lett.*, 63:105, 1989.

[21] B.Efron and R.Tsibirani. Bootstrap methods for Standard Errors, Confidence Intervals, and Other Measures of Statistical Accuracy. *Statistical Science*, 1(1):54–77, 1986.

151

[22] J. D. Farmer and J. J. Sidorowich. Predicting chaotic time series. *Phys. Rev. Lett.*, 59(8):845–848, 1987.

[23] J.D. Farmer and J.J. Sidorowich. Exploiting chaos to predict the future and reduce noise. In Y.C. Lee, editor, *Evolution, Learning and Cognition*. World Scientific, 1988.

[24] J. D. Farmer and J. J. Sidorowich. Optimal shadowing and noise reduction. *Physica* , 47D:373, 1991.

[25] A.M. Fraser. Reconstructing attractors from scalar time series: A comparison of singular system and redundancy criteria. *Physica* , 34D, 1989.

[26] A.M. Fraser and H.L. Swinney. Independent coordinates for strange attractors from mutual information. *Phys. Rev.*, 33A:1134–1140, 1986.

[27] J. Geweke. Inference and forecasting for chaotic nonlinear time series. Technical report, Duke University, 1989.

[28] M. Gorman and K. Robbins. Real-time identification of flame dynamics. In J.Kim and J.Stringer, editors, *EPRI workshop on Applications of chaos* 1991 to appear.

[29] G. Gunaratne, P. Linsay and M. Vinson. Chaos beyond onset: A comparison of theory and experiment. *Phys. Rev. Lett.* 63:1, 1989.

[30] H. Haucke and R. Ecke. Mode locking and chaos in Rayleigh-Benard convection. *Physica* 25D:307, 1987.

[31] X. He. Personal communication.

[32] N.F. Hunter. Application of nonlinear time series models to driven systems. In M. Casdagli and S. Eubank, editors, *Nonlinear Prediction and Modeling*. Addison-Wesley, 1991. to appear.

[33] B. LeBaron. Nonlinear Forecasts for the S&P Stock Index. In M. Casdagli and S. Eubank, editors, *Nonlinear Prediction and Modeling*. Addison-Wesley, 1991. to appear.

[34] P. Linsay. An efficient method of forecasting time series using linear interpolation. *Phys. Lett. A* 153:353, 1991.

[35] G. Mayer-Kress, editor. *Dimensions and Entropies in Chaotic Systems – Quantification of Complex Behavior*, volume 32 of *Springer Series in Synergetics*. Springer-Verlag, Berlin, 1986.

[36] A. Osborne and A. Provenzale. Finite correlation dimension for stochastic systems with power-law spectra. *Physica* , 35D:357-381, 1989.

[37] N.H. Packard. A genetic learning algorithm for the analysis of complex data. *Preprint, University of Illinois*, 1989.

[38] M. Priestley. Spectral analysis and time series. p728 Academic Press, 1981.

[39] T.Sauer, J.A.Yorke and M.Casdagli. Embedology. *Submitted to Comm. Math. Phys.*, 1990.

[40] J.D. Scargle, Modeling chaotic and random processes. *To appear in Journal of Astrophysics*, 1990.

[41] G. Sugihara and R. May. Nonlinear forecasting as a way of distinguishing chaos from measurement error in a data series. *Nature*, 344:734–741, 1990.

[42] F. Takens. Detecting strange attractors in fluid turbulence. In D. Rand and L.-S. Young, editors, *Dynamical Systems and Turbulence*, Berlin, 1981. Springer-Verlag.

[43] J. Theiler. Spurious dimensions from correlation algorithms applied to limited time series data. *Phys. Rev.*, 34A:2427-2432, 1986.

[44] J. Theiler. Estimating fractal dimension. *J. Opt. Soc. Am. A*, 7:1055–1073, 1990.

[45] J. Theiler. Personal communication.

[46] H. Tong. Nonlinear Time Series Analysis: A Dynamical Systems Approach. Oxford University Press, 1990.

[47] H. Tong and K.S. Lim. Threshold autoregression, limit cycles and cyclical data. *Journal of the Royal Statistical Society B*, 42(3):245–292, 1980.

[48] B. Townshend. Nonlinear prediction of speech signals. *to appear in IEEE Transactions on Acoustics, Speech, and Signal Processing*, 1990.

[49] A. Weigend, B.Huberman and D.Rumelhart. Predicting the future: a connectionist approach. International Journal of Neural Systems, 1:193-209, 1990.

[50] A. Weigend, B.Huberman and D.Rumelhart. Predicting sunspots and exchange rates with connectionist networks. In M. Casdagli and S. Eubank, editors, *Nonlinear Prediction and Modeling*. Addison-Wesley, 1991. to appear.

[51] G. U. Yule. *Philos. Trans. Roy. Soc. London A*, 226:267, 1927.

[52] D. Zambella and P. Grassberger. Complexity of forecasting in a class of simple models. *Complex Systems*, 2:269, 1988.

Nonlinear Dynamics and Chaos in Arms Race Models

G. Mayer-Kress

We discuss several nonlinear generalizations of the classical arms race models of L.F. Richardson. We consider models with discrete time evolutions and economic constraints. For two nation models one can find chaotic solutions in perfectly symmetric cases. For three nation models and the possibility of alliance formation, we can find large stable regions for simple approaches to arms-control fixed points separated by control surfaces at which critical changes of alliances occur. In a more specific arms race model which involves three different strategic weapons systems we discuss different types of stable and unstable solutions of scenarios related to the introduction of strategic defense (SDI) systems. Armsraces between two nations in which decisions are based on the power distribution within each of the nations are simulated in a discrete generalization of Kadyrov models of two coupled cusp control surfaces.

We include a brief discussion of the use of stochastic forcing modelled by Chapman-Kolmogorov equations to estimate robustness of the models against incomplete information and external perturbations. For an efficient search for specific classes of solutions we discuss an example of the application of genetic algorithms to find solutions with balance of power for the discrete three nation Richardson model. Finally we make some speculation about how these models can be seen in a more general context of models related to the human impact of global change problems.

Mathematics Department, University of California at Santa Cruz, The Santa Fe Institute, 1120 Canyon Rd, Santa Fe, NM, Center for Nonlinear Studies, Los Alamos National Laboratory, gmk@goshawk.lanl.gov

1. Introduction

We analyze computational models related to international security problems from a perspective of nonlinear, stochastic dynamical systems with discrete and continuous time evolution. The arms race models we consider are a special class which is related to population dynamics and which was first introduced by L.F. Richardson after WW I [27]. The examples we discuss, however, have discrete time dependence. For certain ranges of their control parameters, some of these models exhibit deterministic chaos [16], [28], and we discuss how this behavior limits our ability to anticipate and predict the outcomes of various situations [1]

In a generalization of a discrete version of Richardson's equations we model the competition among three nations [24], the two weaker ones will form an alliance against the strongest one until the balance of power shifts (see e.g. [34], [2], [7], [18] for related discussions). The alliance formation factor and economical constraints in the model introduce non- linearities which cause multiple stable solutions, bifurcations between fixed points solutions and time dependent attractors. We have identified parameter domains for which the attractors become chaotic. This fact can be interpreted as an irregular sequence of changes of alliance configurations.

The attractors of a dynamical systems model is of limited validity if it is not robust against finite external perturbations, the magnitude of which is estimated from the context. Instead of attempting to model specific types of external perturbations, we model them by a sequence of uniformly distributed, bounded random numbers. Thus we represent limited information about the values of the variables in the model. We observe that different solutions of the model can have very large differences in their sensitivity to external perturbations. We interpret strong sensitivity of the solutions as an indicator for deterministic chaotic (possibly transient) dynamics (see e.g. [29]).

For each point on a solution curve we can define a distance to a perfect "balance of power" for which the arms expenditures of the two allies balance the expenditures of the strongest nation. In [12] we used genetic algorithms to search for sub-manifolds in the 15-dimensional parameter space of the model, for which there exit non-trivial solutions with perfect balance of power. We found evidence that those solutions are not rare and that they correspond to arms-control configurations which are globally stable.

In democratic societies the decision of the government on international relations is reflecting the public opinion. In a simple model Kadyrov [19] describes the formation of public opinion in two competing nations: The

---

[1] See e.g.[3] ,[4] ,[8] ,[9] ,[10] ,[30] for a general introduction to chaos.

population is divided into "hawks" and "doves" and the influence of each of the groups on their own government will determine the strength of the corresponding configuration in the competing nation. The structure of this model correspond to two coupled cusp systems. We have used computational models to unfold the bifurcations in this model [1]. We have studied both continuous time and discrete time models and found large regions in parameter space, for which the solutions and the dynamics of the two classes of models are equivalent. Furthermore we were able to extend the range of the discrete model and identify parameter domains for which multiple periodic solutions and also chaotic solutions become stable.

2. Arms Race Models in Discrete Time

The pioneering work of Richardson [27] set the stage for subsequent attempts to analyze quantitatively many questions of strategic military and economic competition between – and among – nations. In this section we discuss several simple and discrete variants of Richardson's equations [24], primarily as a brief exercise to indicate how one goes about formulating and analyzing nonlinear models of socio-political issues. We recall that the original Richardson equations [27] as applied to the arms race between two nations have two variables evolving in continuous time. This assumption, standard and justified in physical models, leads to extremely restrictive theoretical consequences: Solutions to the original Richardson's equations can not exhibit more complex behavior than fixed points or limit cycles. However, if one considers discrete time – representing, for example, the fact that budget decisions are typically approved on an annual basis, relevant information is not provided continuously – and adds nonlinear terms to represent limited economic resources, already in the case of two nations one can observe all the striking features found in the nonlinear dynamical systems discussed above. We should perhaps stress at this point that in our opinion the intention of using simple, nonlinear dynamical models is different from the ones of classical game theory where one was interested in finding an optimal strategy as the unique solution to a given problem. One of the lessons of chaos theory is that individual solutions (histories) are basically non-reproducible and therefore of very limited relevance. A system might have many different but equivalent solutions to the same problem.

2.1 2-Nation Richardson Models

In order to present clarify the process of constructing a simple discrete time Richardson model, let us consider the following example of two competing nations: Let $x_\alpha(t)$ be variables describing the arms expenditures of nation

α , $\alpha = 1, 2, ..., d$ denote the components of the system (i.e. the players) at time t. The dimension d can be small, i.e., $d = 2$ if NATO and WTO are considered or the USA and the USSR. If Europe and China are taken additionally and independently one has $d = 4$. In the various local conflicts other countries are involved, but the dominant variables are always only a few. In physics the $x_\alpha(t)$ are often called "order parameters" or "modes" or "amplitudes".

The stable stationary states x_α^* depend, on certain control parameters a which in this context express policies, strategies, economical constraints, etc. We usually first explore how the steady state $x_\alpha^*(a)$ depends on a and then alter the control parameters to adjust the state according to our demands and aims (wishes).

2.1.1 Example 1: Logistic Model

The appearance of chaotic solutions has been described in a discrete Richardson type model [28]. There the model is given by:

$$x_{t+1} = 4ay_t(1 - y_t) \equiv f_a(y_t)$$

$$\tag{1}$$

$$y_{t+1} = 4bx_t(1 - x_t) \equiv f_b(x_t)$$

where $x = x_1$ and $y = x_2$ in the above notation and denote the respective fractions of the available resources which countries X and Y devote to armament.

It seems quite reasonable to assume a country's armament x_{t+1} to be determined by the hostile armament y_t of the previous year; less reasonable is the factor $(1 - y_t)$, since it reduces the own efforts to 0 if the hostile country has the largest possible $(y_t = 1)$ armament.

The model mentioned above is *basically one*-dimensional with two years steps, $x_{t+2} = f_a(f_b(x_t)) \equiv f_{a,b}(x_t)$, correspondingly for the other country with $f_{b,a}(y_t)$. Both countries are simultaneously in a steady state $x^*, y^* (= f_b(x^*))$, simultaneously bifurcate to a periodic state, or together enter chaos.

In [28] the transition to chaos was associated with unpredictable behavior, crisis-unstable arms races, and therefore with an increased risk for the outbreak of war Although this interpretation can be disputed (see e.g. the discussion in [18], [16], it has become clear that the possibility of transitions to chaotic behavior has to be taken into account in arms race models. A second interpretation of the role of chaos in arms race models has been discussed in [16]: Bounded, small-scale chaos can be part of a stable arms control situation. There the other side's response cannot be anticipated in detail (possibly because of internal political problems of one country), but the stability of the attractor itself allows for confidence in the fact that no disastrous surprises will occur. This would be expected if there exists a basic consensus within one nation about defense policy, but details of the budget are subject to internal discussions. The effect of couplings between internal opinion formation (between "hawks" and "doves") and interactions between two nations has been discussed in [1].

The appearance of this type of small scale chaos has to be distinguished from large chaotic fluctuations which would lead to configurations in which "crises" (see e.g. [15] for a technical definition and discussion) which could lead to unbounded arms races, war, or economic collapse.

2.1.2 Example 2: Resource Limited Response

A similar arms race model for which the nonlinearity can be interpreted as economic constraint has been discussed in [16].

It is given by:

$$x_{t+1} = x_t - k_{11}(x_t - x_s) + k_{12}y_t(1 - x_t)$$

$$y_{t+1} = y_t - k_{22}(y_t - y_s) + k_{21}x_t(1 - y_t)$$

(2)

x_t and y_t again denote the fraction of the available resources that is devoted to armament in the countries X and Y during the year t. The *change* of armament or military expenditure, $x_{t+1} - x_t$, is taken to be proportional to the military effort y_t of the other country. The rate factor was considered as constant by Richardson. We assume, instead, that this factor itself depends on the still available resources and put $k_{12}(x_m - x_t)$. The largest available fraction of resources, x_m, can be taken as 1 by rescaling the equation of motion. k_{12} and k_{21} (positive) correspond to the defense intensity of Richardson. x_s, y_s are the countries "natural" self-establishing armament levels, corresponding to Richardson's grievance. k_{11} and k_{22} are the (positive) expense coefficients forcing the approach to x_s, y_s, were there not the hostile country.

In the case that both countries are alike and we choose $x_s = y_s \approx 0.1$, $k_{11} = k_{22} \equiv \kappa \approx 1.4$, and $k_{12} = k_{21} \equiv k$. If both countries start initially at the same level $x_0 = y_0$ then $x_t = y_t$ for all times t, and the model equations can be simplified to $x_{t+1} = f(x_t)$. Now f is a quadratic function of x_t. For small k one has a stable fixed point; increasing the defense intensity k

leads to period doubling and that whole scenario preceding chaos. At about $k > 3$ chaos sets in, the threshold depending, of course, on the self-expense-coefficient κ. This situation is shown in fig. 1 for $k = 3.6$:

The model is simulated and the results displayed in a modern spread-sheet [2]. We have windows for the equations, numerical values of the simulation, representation of the timeseries as well as a phase space plot, here in the symmetric case we display x_{t+1} vs. x_t. The attractor consists of two bands between the system will oscillate periodically, i.e. every other time step the system will in a domain around $x = 0.4$ and then it will be mapped into a region around $x = 0.8$. The uncertainty about the expected state of the system is about 0.3 in the first band and about 0.1 in the second case.

This model can be considered as a nonlinear, discretised version of Richardson's equations. It is known that both nonlinearity and discretisation can lead to instabilities and ultimately to chaos. This is an effect of overshooting, but evidently present in arms race reality due - among other reasons - to the discontinuous data flow and political decision mechanisms. Thus we want to stress the fact that the discretisation is done on the modeling level and is not supposed to reproduce the behavior of the continuous equations. The corresponding continuous version of the model would typically describe quite a different situation. Similar problems of interpretation of discretisation effects in the context of Lanchester models of combat have been discussed in [31].

Some surprising features of the above resource-limited model are that in general it shows a *sharp* transition to instability with no preceding period doubling which signals the threatening instability, (interpreted as the outbreak of war or a break-down of the model) and that the *threshold to instability is much lower*, if both countries are more independent and not unrealistically assumed to react in precisely the same way.

2.1.3 Example 3: Global Resource Limitation

A slight variation of this model with significantly different properties is given if the resource limitation affect all arms expenditures. It is given by:

$$x_{t+1} = x_t + \left[-k_{11}(x_t - x_s) + k_{12}y_t\right](x_m - x_t)$$

$$\tag{3}$$

$$y_{t+1} = y_t + \left[-k_{22}(y_t - y_s) + k_{21}x_t\right](y_m - y_t)$$

---

[2] This class of models has been implemented by Tom Affinito, UCSC

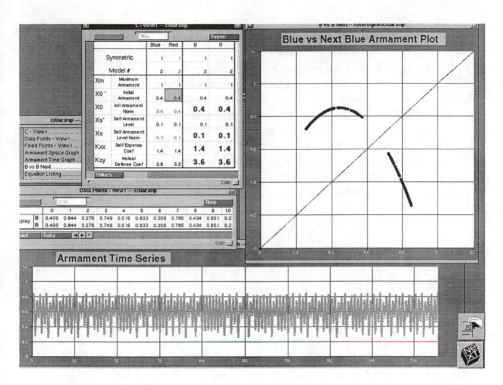

Figure 1: Spreadsheet interface for 2-nation discrete Richardson models

Screen array of the spread-sheet interface for simulation of 2-nation Richardson model. The five windows contain information about the elements of the spreadsheet, parameters and initial conditions, numerical values of the time series, graphical display of the time series, phase-space plot of the solutions. Not shown is the window with model equations and voice annotation.

This simple global resource limited model seems to behave according to our expectations: If the mutual defense intensity k is small, reflecting confidence into the other country's peacefulness, there is a rather low level military expenditure, only slightly above the self-caused level s, increasing with k. For the case of two countries of equal abilities: $x_s = y_s =: s$, $k_{11} = k_{22} =: \kappa$, and $k_{12} = k_{21} =: k$ the fixed points of the system are given by $P_1 = (1,1), P_2 = (1, s + \frac{k}{\kappa}), P_3 = (s + \frac{k}{\kappa}, 1)$, and $P_4 = (\frac{\kappa s}{\kappa - k}, \frac{\kappa s}{\kappa - k})$.

Of course, the P_i have a meaning only if they belong to the unit square $0 \leq x, y \leq 1$. Hence P_1 always is meaningful, namely the state of maximum military effort of both countries. $P_{2,3,4}$ either are meaningful altogether or none. The condition for being allowed states (within the unit square) reads

$$k \leq k_c , \quad k_c := \kappa(1 - s). \tag{4}$$

At k_c all stationary states coincide. P_4 is on the angle bisector between the self-armament (s, s) and maximum armament $(1, 1)$, while P_2, P_3 are on the right or upper edge.

Are these steady states attained? This is decided by linear stability analysis. The (only) eigenvalue at P_1 is $\lambda^{(1)} = 1 + k_c - k$. Any (negative) deviation $(\delta x_t, \delta y_t)$ from P_1 shrinks or grows by the factor $\lambda^{(1)}$. Hence, if $k > k_c$ the state of maximum armament is stable, unless k exceeds $2 + k_c$. The other states P_i are not only meaningless in this range but also turn out to be unstable.

P_4 is an attracting stationary state if and only if it is meaningful at all, i.e., for $k < k_c$. It has two eigenvalues $\lambda_{1,2}^{(4)}$ and two attracting eigendirections, while $P_{2,3}$ have always for $k < k_c$ as well as $k > k_c$ one attracting and one repelling eigendirection, i.e., they are never attained under realistic (fluctuating) conditions.

If k exceeds a threshold value k_c (range $k_c \leq k < 2 + k_c$) the countries spend the highest possible expenditure for arms, irrespective of their internal demands. Arms race is unavoidable if the mutual defense factor is not below k_c. The change from the low level armament to the high level one occurs continuously with k.

2.2 3-Nation Richardson Model

Let us focus here on an extension of the Richardson equations to model an arms race among three nations; this introduces the additional complication of possible alliance formations (see for example [34]). With three nations the natural choice for an alliance is between the two weaker nations, as otherwise the two nations in the alliance would have dominant superiority over the third one and the competition could be reduced to that between the nations within the alliance. For the model, let us therefore assume that we have a situation in which the nations X, Y, Z have normalized arms expenditures amounting to values x_n, y_n, z_n, where the index n indicates a discrete time unit and corresponds to a typical decision period : for example, x_n can indicate the military budget in year n. We study the generic situation in which we can assume without loss generality: $x_n > y_n > z_n$. In this case, by our assumptions Y and Z would form an alliance against X. The equations describing the armament level in the year $n+1$ contain, for each nation, four relevant factors: (1) the armament level in the previous year (x_n, y_n, z_n); (2) the intrinsic self-armament level that each nation wants to acquire, independent of external threats (x_s, y_s, z_s); (3) the external threat from hostile nations; and (4) the economical limitations (x_m, y_m, z_m). The terms in the equations describing the external threats are different for nations who have to defend themselves against an alliance and those who are in an alliance. In the first case the threat consists of the sum of the expenditures of the adversaries, $(y_n + z_n)$, whereas in the second case the threat from the adversary X is reduced by the expenditures of the ally $(x_n - z_n)$ for Y, $(x_n - y_n)$ for Z.

We regard the arms expenditures – the x_n, y_n, z_n – as the variables whose time evolution we want to follow and the other variables – x_s, y_s, z_s and x_m, y_m, z_m – as parameters, analogous to the carrying capacity N_∞ in the simple logistic population dynamics. Pursuing this analogy further, we see that to couple the variables together we need the analog of the fecundity parameter, r. Here there are several parameters, which effectively represent the results of political decisions about rates at which goals are to be approached. The parameters (k_{11}, k_{22}, k_{33}) determine how fast each nation tries to achieve its intrinsic armament level independent of external threats, while the parameters (k_{23}, k_{13}, k_{12}) describe the rate at which a perceived external threat is countered. For the condition that $x_n > y_n > z_n$, the resulting discrete are given by:

$$
\begin{aligned}
x_{n+1} &= x_n + \left[k_{11}(x_s - x_n) + k_{23}(y_n + z_n) \right](x_m - x_n) \\
y_{n+1} &= y_n + \left[k_{22}(y_s - y_n) + k_{13}(x_n - z_n) \right](y_m - y_n) \qquad (5) \\
z_{n+1} &= z_n + \left[k_{33}(z_s - z_n) + k_{12}(x_n - y_n) \right](z_m - z_n)
\end{aligned}
$$

If in the course of the simulation, the ordering in the arms expenditures is changed, the equations are rearranged in such a way that a new alliance is formed among the new minor powers. In view both of the large number of parameters and of the uncertainty in their individual values, one should study [24] the equations for a wide range of parameters and attempt to isolate those domains of parameters for which the solutions show robust, regular behavior from those in which sensitive, chaotic behavior is obtained.

A preliminary analysis [24] of typical solutions of this model shows (unsurprisingly) multiple attractors, separated by surfaces forming boundaries of the basins of attraction. There are indications for small scale chaos in several regions of state space for certain combinations of parameters. It appears also that close to boundaries in parameter space at which transitions in the alliance structures take place, the system becomes (again, unsurprisingly) very sensitive to external noise (see Fig. 2).

This phenomenon could be helpful in the computational identification of crisis domains. Although no one could take the conclusions drawn from such an over-simplified model as the sole basis for any political decision, we feel that the attempt to understand and to quantify various causal relationships will lead to increasingly sophisticated models for specific aspects of these problems; when such models are produced, one will certainly need the insight gained from simpler models and the techniques of modern nonlinear dynamical systems theory to analyze these models.

2.3 Strategic Defense Initiative (SDI) Model

A more sfigpecific arms race model, based on the same mathematical tools of discrete time dynamics, has been applied to model the impact of strategic defensive systems (SDI) on the superpower arms race [29]. This simple model only includes three S.D.I. elements: (a) intercontinental ballistic missiles (ICBMs), the component of offensive, strategic, nuclear warfare; (b) anti-ICBM satellites, designated to attack and destroy ICBMs from space; and (c) anti-satellite missiles or other weapons launched to destroy overhead anti-ICBM satellites before these satellites can destroy the ICBMs about to be launched.

In order to define the model we have to introduce a set of parameters: The index i denotes the superpowers, $i = 1$ for US, $i = 2$ for SU. Let μ_i be the complete capital cost per ICBM for side i (including silos and other support apparatus; for simplicity, we include all lifetime operating costs of the weapon system in its capital costs.) and let $M_n(i)$ represent the number of such ICBMs held by side i at year n of the arms procurement race; then $\mu_i M_n(i)$ is the total cost of offensive missiles to side i since the "beginning" of the arms race and $\mu_i(M_n(i) - M_{n-1}(i))$ is the offensive missile cost during year n. Similarly, let σ_i be the cost per anti-ICBM satellite, $S_n(i)$ be the number of such satellites, α_i be the cost per satellite-killer weapon, and $A_n(i)$ the number of such killers of the anti-ICBM satellites. The cost per

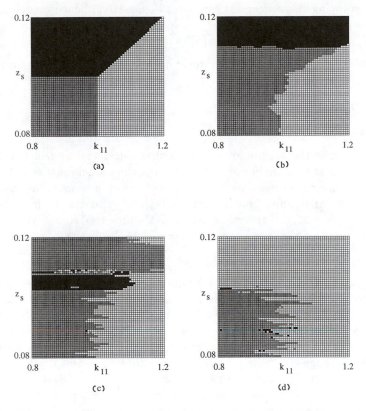

Figure 2: Effect of Noise Perturbation on Typical Parameter Settings

A 2-dimensional slice through the twelve dimensional parameter space of the model through the standard parameter set. The slice consist of $64 \times 64 = 4096$ time histories over $T = 33$ time steps (years). The x-y coordinates correspond to the range of values for parameters k_{11} (the x-axis) and z_s (the y-axis). The grey coding corresponds to the relative strength of different countries: bright $= X$ is strongest (i.e. $x_t > y_t$ and $x_t > z_t$), medium $= Y$ is strongest, dark $= Z$ is strongest after 33 time steps. The boundary between regions of different shades of grey corresponds to a change in the alliance configuration, a case that we associate with a major crisis. (a) no noise $\sigma = 0$ Effect of low level perturbations in the system parameters on a slice in the k_{11}, z_s plane can be seen in:(b) noise, $\sigma = 0.001$ averaged over $N = 1000$ time histories, (c) $\sigma = 0.003$, $N = 1000$ (d)$\sigma = 0.005$, $N = 1000$

164

year of deploying these weapons can never be negative; hence we define and use [3]

$$\hat{\theta}(x) = x \; \theta(x) = max(x,0)$$

Then the total expenditure for strategic nuclear warfare, during year n, by side i, is

$$D_n(i) = \mu_i\hat{\theta}(M_n(i) - M_{n-1}(i)) + \sigma_i\hat{\theta}(S_n(i) - S_{n-1}(i))$$
$$+\alpha_i\hat{\theta}(A_n(i) - A_{n-1}(i)) \quad (6)$$

If D_i represents the maximum resources that side i can or will devote to strategic nuclear warfare in any one year, then the model must require that $D_i - D_n(i)$ never be negative. In fact we expect a smooth decrease in strategic nuclear weapon procurement as $D_n(i)$ approaches D_i from below. Thus $M_{n+1}(i) - M_n(i)$, the number of ICBMs procured in one year at the end of n years of the arms race, is assumed to be limited by the factor $\hat{\theta}(\frac{D_i-D_n(i)}{D_i})$.

To satisfy their present strategic functions, the ICBMs of side i must deliver to side j's territory a given minimal number of warheads which we designate as R_i. If the strategic function of the ICBMs for both sides is deterrence, as we assume in this model [4], then R_i is the minimum number of re-entry vehicles, belonging to i, which, upon impact on j, produces "intolerable" damage to j; it is usually assumed that $R_i = 500, i = 1$ or 2, i.e., 500 nuclear warheads impacting either the U.S. or the USSR would produce a deterring intolerable damage to the recipient society. (Alternatively, for a first-strike strategy in which the goal is to destroy all of the opponent's ICBMs ("damage limiting" strike), one would have $R_i = M_n(j)$.) Let $R_n(i)$ be the number of retaliatory warheads that i believes will actually get through j's defensive system and strike their designated targets. Then, if $R_n(i)$ is less than R_i, i will perceive that the retaliation task set for their offensive strategic forces will not be accomplished; they will proceed to build more ICBMs within the constraints of economics and resources. If $R_n(i)$ exceeds R_i, i has more than enough to accomplish their purposes. Thus we write the "Missile Builder's Recurrence Relation" as follows (with the possibility of build down ($M_{n+1}(i) < M_n(i)$)):

$$M_{n+1}(i) = M_n(i) + f_i[R_i - R_n(i)]\hat{\theta}\left(\frac{D_i - D_n(i)}{D_i}\right) \quad (7)$$

Let m_i be the number of warheads per ICBM for side i (the mean "MIRV" number), $\eta_n(i)$ the number of i's missiles that are destroyed in their silos

[3] here $\theta(x)$ is the Heavyside unit step function ($\theta(x) = 1$ for $x \geq 0, \theta(x) = 0$ for $x < 0$).
[4] Note that this assumption is based on what has been discussed in the public, not what is actual operating strategy.

by j's first strike ICBMs, and $\Delta M_n(i)$ the number of i's retaliatory missiles destroyed in their boost phase by j's satellites. (Unless destroyed or intercepted, all missiles are assumed to work perfectly; there are no launch or flight failures. If perfect accuracy and effectiveness is not assumed, e.g., if two warheads are assigned to each target, then m_i would be one-half of the mean number of warheads per ICBM). Then the number of retaliatory warheads is:

$$R_n(i) = m_i[M_n(i) - \eta_n(i) - \Delta M_n(i)]\theta(R_n(i)) \tag{8}$$

where the step function makes sure that $R_n(i)$ is a positive number (or zero). We assume that ICBM accuracy has improved to the point that a warhead that gets through the opposing defenses will destroy its target silo with certainty. Thus $\eta_n(i)$ is equal to the number of j's warheads that get through i's satellite defense, i.e.,

$$\eta_n(i) = m_i[M_n(j) - Q_n(j) - \Delta M_n(j)]\theta(\eta_n(i)) \tag{9}$$

where $Q_n(j)$ is the number of ICBMs that j does not fire in its first strike against the i silos, keeping them in reserve for a second strike. The number held in reserve must be sufficient for the deterrent purposes to be fulfilled after the opponent's satellites have taken their toll of offensive missiles:

$$Q_n(j) = \frac{1}{m_j}R_j + \Delta M_n(j) \tag{10}$$

The number of j missiles killed in their boost phase by i's defensive satellites depends upon the number of i satellites over j's territory at any instant. This latter number, in turn, depends upon the total effective number of i anti-ICBM satellites in orbit, which we designate as $S_n^{eff}(i)$. In [5] $\Delta M_n(j)$ is estimated as:

$$\Delta M_n(j) = \left[\frac{1}{\beta_i}S_n^{eff}(i)\right]^2 \tag{11}$$

The parameter β_i is given in terms of satellite "brightness", target missile "hardness", and geometric factors; it takes into account the fact that not all satellites can be brought into battle at anyone time: because they move in orbits, only a fraction (determined geometrically in [6]) is within destructive range of the ICBM fields at any moment.

The minimum number of satellites required to do defend against surprise attack occurs when $S_n^{eff}(i) = \beta_i\sqrt{\Delta M_i}$. If $S_n^{eff}(i)$ exceeds this number, i could build down some of the excess satellites. Hence, if $S_n(i)$ designates the actual number of missile-killing satellites possessed by side i in year n of the arms race, satellites will be procured according to the "satellites builders recursion relation":

$$S_{n+1}(i) = S_n(i) + g_i[\beta_i \sqrt{M_n(j)} - S_n^{eff}(i)]\hat{\theta}\left(\frac{D_i - D_n(i)}{D_i}\right) \tag{12}$$

where g_i is another pair of free, positive, model parameters that correspond to a rate at which ABM satellites will be built. This pair of recursion relations is not completely specified until $S_n^{eff}(i)$ is related to $S_n(i)$.

The effective number of satellites, those that can actually destroy rising ICBMs, (remembering that, via the parameter β_i, only a fixed fraction of the effective number of satellites is actually capable of doing any ICBM killing at any moment) is less than the procured number since some of them may be destroyed at the beginning of the battle, before they have had a chance to be brought into use, by the aggressors anti-satellite weapons, which number $A_n(j)$. If these weapons are also assumed to have perfect accuracy, the number of satellites destroyed equals the number of weapons fired at them. Thus

$$S_n^{eff}(i) = [S_n(i) - A_n(j)]\theta(S_n(i) - A_n(j)) \tag{13}$$

Suppose that $Z_n(j)$ is the maximum number of j satellites that will allow i to penetrate the j defense to the minimal desired degree. If the effective number of j's satellites is greater than this number, then i will have to destroy more, implying the procurement of more satellite killing missiles, if allowed by resource constraints. Hence the "satellite-killer-missile builder's recursion relation" is

$$A_{n+1}(i) = A_n(i) + h_i\, [S_n^{eff}(j) - Z_n(j)]\, \hat{\theta}\left(\frac{D_i - D_n(i)}{D_i}\right) \tag{14}$$

where the h_i determine the rate at which ASAT systems are deployed. We thus have three pairs of procurement recursion relations all six relations being coupled to each other.

All that remains is the specification of $Z_n(j)$, which determines the number $[\frac{Z_n(j)}{\beta_j}]^2$ of i's ICBMs which can be destroyed. In addition, $\eta_n(i)$ of i's potentially retaliatory missiles are destroyed in their silos by j's presumed first strike. Thus in order for i to be able to deter j's first strike, e.g., to be sure that R_i of i's warheads will get through after the first strike, $Z_n(j)$ is limited by

$$R_i \geq m_i\left[M_n(i) - \eta_n(i) - \left(\frac{Z_n(j)}{\beta_j}\right)^2\right] \tag{15}$$

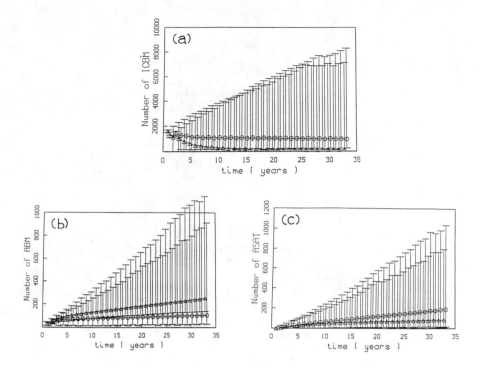

Figure 3: Weapons Expenditure: Accelerated Deployment

Temporal evolution of the number of strategic weapons of the US (o) and of the SU (△) for accelerated deployment of SDI system (see [29] for numerical values for the system parameters). (a) Intercontinental ballistic missiles (ICBM) (b) SDI satellite systems (ABM) (c) Antisatellite weapons (ASAT). The errorbars indicate minimal and maximal values of the simulation under 10^5 runs of the model under the influence of 50% random perturbations of the build up parameters at each timestep. Note the wide range of possible outcomes in spite of a decreasing average expenditure.

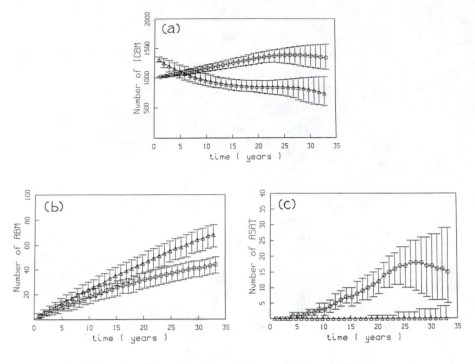

Figure 4: Weapons Expenditure: Reduced MIRVing

Same as in Fig. 3 with reduced build up parameters and number of warheads per missile reduced by factor of 2.

Substituting for $\eta_n(i)$, the limiting value for $Z_n(j)$ is determined by

$$\left[\frac{Z_n(j)}{\beta_j}\right]^2 \geq M_n(i) - \frac{R_i}{m_i} - m_j\left(M_n(j) - \frac{R_j}{m_j} - 2\left[\frac{S_n^{eff}(i)}{\beta_i}\right]^2\right) \qquad (16)$$

The six coupled non-linear procurement recursion relations can be solved for any n given the starting values,[5]

$$M_0(i), S_0(i), A_0(i), M_{-1}(i), S_{-1}(i), A_{-1}(i)$$

the deterrence parameters R_i, the ICBM and satellite characteristic parameters m_i and β_i, the economic constraint and cost parameters $D_i, \mu_i, \sigma_i, \alpha_i$, and the procurement model parameters f_i, g_i, and h_i (a total of $2 \times 13 = 26$ required numbers).

The results of that idealized model indicate that for most parameter combinations, the introduction of SDI systems leads to an extension of the offensive arms race rather than a transition to a defense dominated strategic configuration. For the case of a strongly accelerated arms build up (either offensive or defensive) we observe a loss of stability of the solutions that we interpret as a transition to unpredictable chaos (See fig.3). Computationally this is observed by iterating the model equations under the influence of stochastic perturbations (applied at every time step) over a given period of time and then determining the range of observed outcomes from the simulation. The stochastic nature of these perturbations is very important for reliable robustness estimates: in contrast to linear systems, perturbations in individual variables or parameters can not be used to make inferences on the system's behavior under simultaneous perturbations of many different factors. Murphy's law pays respect to this phenomenon: it often seems that many disadvantageous factors in a complex system "conspire" to achieve the worst possible outcome.

A reduction in the number of offensive weapons, i.e., an approach to a defense dominated strategy, was observed if either the number of reentry vehicles per ICBM (MIRV) is limited to much smaller values than presently realized or if the accuracy of offensive weapons is significantly reduced (See fig.4).

2.4 Kadyrov models

In this class of models interactions between influential groups of different political opinions ("hawks" and "doves") are incorporated. The relative

---

[5] Note that we have to specify initial conditions for two successive time steps because of the inherent time delay in the system. The dynamical equations only come into action for positive times.

strength of each group in each country not only determines the armsrace efforts of the nation but also the changing influence of the political groups in the other nation. In the case of the strategic arms race between the US and the USSR, the interactions between the supporters of Gorbachov, his foreign policy and its influence on, but also from western peace movements might serve as a simplistic analogy. In [1] we were interested in general mathematical structures of Kadyrov type models. Similar to the case of the previous examples of Richardson models, we discussed not only the original continuous time model, but also discretized versions thereof. The discrete models have the advantage of higher speed and accuracy in computational implementation and realization. [6]

The degree of relative political influence of hawks and doves in each of the societies depends in this model on two control parameters: the perceived cost of a war, b, and the fear of defeat, a. The state variable, x corresponds to the governmental policy, ranging from peaceful ($x < 0$) to hostile ($x > 0$). A detailed discussion of the continuous and discrete version of this model is contained in [1]. Here we want to review some of the properties of the discrete version.

In order to model a single cusp we define for a, $b \in \mathbf{R}$, $b > 0$ the family $f(x, a, b) = tanh(bx) + a$ (see. Fig. 5.)

We have a splitting parameter, b, and a normal parameter, a. For $b < 1$, the slope of $f(x, a, b)$ is everywhere less than one, and therefore there exists only one fixed point, x_c (that is, $f(x_c, a, b) = x_c$), and this is always stable and globally attracting. The location of x_c is determined by the values of both a and b.

Now increase b, keeping $a = 0$. Since $f'(0, a, b) = b$, the fixed point x_c becomes unstable for $b = 1$. Simultaneously, two new fixed points ($x_l <= x_c <= x_r$) are created. They are stable since $f'(x, a, b) - > 0$ for $x - > \pm\infty$, and f is invertible. The degeneracy of the simultaneous creation of x_l and x_r can be unfolded by allowing $a \neq 0$. The bifurcation from solutions of one fixed point to those with three fixed points is of fold (saddle-node) type, and corresponds to the fold in the ODE case. For any value of $b > 1$ we can find $a_s \neq 0$ such that the system undergoes a saddle-node bifurcation at (a_s, b). We observe that for all $a \in R$, $b > 1$, for which we have three fixed points, x_l and x_r are stable, while x_c is unstable. For $a > 0$ we have $x_c < 0$ and for increasing a, x_l and x_c will coalesce and disappear in a fold bifurcation. Analogously for $a < 0$, x_c and x_r will disappear. From the construction, we see that approach to the stable fixed points x_l and x_r is always monotone, and that ($|f(x, a, b) - x|$) goes to

---

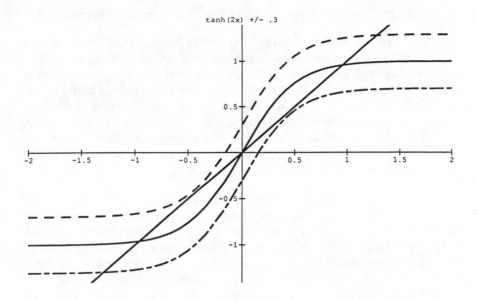

tanh(2x) +/- .3

Figure 5: Discrete Cusp Model: Smooth Map

Graph of $f(x,\ a,\ b)\ =\ tanh(bx) + a$ for $b = 2$ and $a = 0$ (solid), $a = -0.3$ (dot-dashed), $a = 0.3$ (dashed). Note the bifurcations from one stable fixed point (lower branch) to 2 stable fixed points + 1 unstable fixed point to one stable fixed point (upper branch) as parameter a is increased.

infinity for $|x| \rightarrow \pm\infty$ at rate at which the fixed points are approached.

The eigenvalues of the fixed point, i.e. their local stability properties as well as their separation, are determined both by the splitting parameter b and the normal parameter a. We also see that this system has strong symmetry properties, which suggest nongeneric behavior. In order to introduce a controlled and interpretable symmetry-breaking condition for the generic unfolding of the coupled cusp system, and also in order to be able to find the fixed points analytically, we introduce a continuous, piecewise-linear function, $p(x, a, b_L, b_R)$, with a set of solutions equivalent to the hyperbolic tangent map $f(x, a, b)$.

The map $p(x, a, b_L, b_R)$ consists of three linear parts, each determined by parameters specifying slopes and y-intercept, determined by (b_L, b_R), of the local, linear segment. Whereas the relative location of the three pieces with respect to each other controls the bifurcation set, the local slopes permit explicit control over local stability properties related to the fixed points. This factor becomes especially relevant in the coupled case when we not only consider fixed points, but also periodic, quasi-periodic, and potentially chaotic solutions.

The explicit representation of the piecewise-linear mapping is given by

$$x_{n+1} = p(x_n, a, b_L, b_R) = p_o(x_n, b_L, b_R) + a \qquad (17)$$

where

$$p_o(x, b_L, b_R) = \begin{cases} s_2\, x + |b_R| \left(e^{s_1\, b_L} - s_2\right) & , \quad |b_R| < x; \\ e^{s_1 b_L}\, x & , \quad -|b_L| < x < |b_R|; \\ s_3\, x - |b_L| \left(e^{s_1\, b_L} - s_3\right) & , \quad x < -|b_L|. \end{cases} \qquad (18)$$

Here b_L and b_R play the role of the splitting factor, and a plays the role of the normal factor, of the continuous-time model studied above. For a positive values of b_L and a, the mapping is shown in Fig. 6. As the splitting factor, b, changes from a negative to a positive value, it is evident that the central region of this mapping displays the divergence and inaccessibility characteristics of the standard cusp-catastrophe.

We now make the generalization to the two nation model with the state of one nation, x, and the state of the other nation, y. The change of state of both nations is represented by iteration of the following map:

$$\begin{aligned} x_{n+1} &= p_o(x_n, b_L, b_R) + a_1 y_n + a_o, \\ y_{n+1} &= p_o(y_n, d_L, d_R) + c_1 x_n + c_o. \end{aligned} \qquad (19)$$

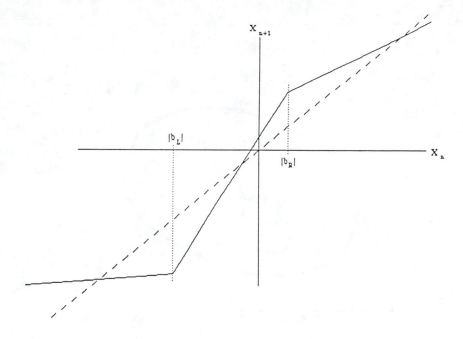

Figure 6: Discrete Cusp Model: Piecewise Linear Map

Plot of the piecewise linear model for $a > 0$ and $b_L > 0$.

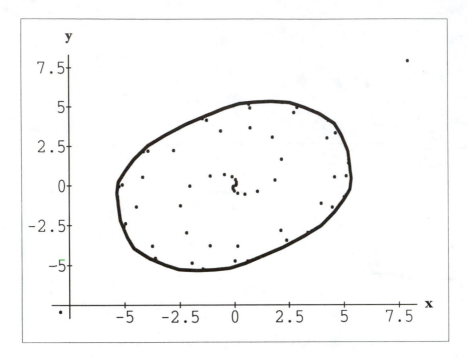

Figure 7: Quasi-Periodic Attractor in Discrete Kadyrov-System

Phase portrait of symmetric case for $a_1 = -0.8$, $c_1 = 0.8$, $b_L = 1.0$ and $d_L = 1.0$, a region displaying a quasi-periodic attractor.

Coupling is achieved through the normal factors. Analytically solving for the fixed points of this mapping, we are able to obtain a tableau of the dependence of the mapping on the control parameters.

Investigating the phase portrait of this mapping for the symmetric case where

$$a_o = c_o = 0, b_R = b_L, d_R = d_L$$

we find much of the same behavior as found in the continuous model.

Increasing the slopes of the discrete mapping from 0.5 to 0.8, the periodic attractor changes to the quasi-periodic attractor of Fig. 7.

In Fig. 8, we were able to find chaos, something not present in the ODE case as guaranteed by the Poincaré-Bendixson theorem.

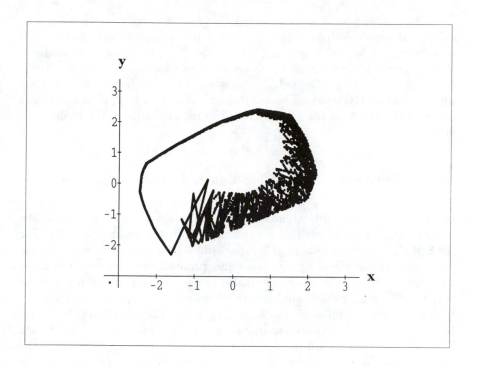

Figure 8: Chaotic Attractor in Discrete Kadyrov-System

The chaotic attractor found in the same region of figure 7 with all slopes equal to 0.5 except for s_3 of nation Y which has a value of -1.5.

The existence of chaos in this case is generated through the local separation of nearby orbits, which is caused by those sections of the function, $p_o(x_n, b_L, b_R)$ with $|p'_o(x_n, b_L, b_R)| > 1$. The global boundedness of the system is guaranteed by those sections of $p_o(x_n, b_L, b_R)$ for which $|p'_o(x_n, b_L, b_R)| < 1$. The balance between these two factors typically results in chaotic attractors. This controlled bifurcation to chaotic attractors is quite different from the appearance of spurious solutions due to discretisation.

3. Sensitivity analysis and Chapman-Kolmogorov equations

In the simple socio-political models discussed above, the estimation of the parameters and initial conditions typically involves large uncertainties. For the SDI model [29] we have used Monte Carlo simulations to estimate the robustness of the solutions against those uncertainties, represented by stochastic forcing [21], [22]. A more systematic method was discussed for discrete maps in [17]: Under the assumption that we have estimates of the size of the (bounded) uncertainties, we can introduce a dynamic equation (Chapman-Kolmogorov equations) for measures defined on the state space of the model. The support of the transformed measures allows us to get an estimate of the range of possible outcomes of the time evolution of the states given by the initial distribution. For example when we apply the Chapman-Kolmogorov equation to a distribution of initial conditions which is concentrated on the (coexisting) attractors of the discrete Kadyrov model we can compute the range of states that the system can reach for a given level of perturbations [20]. For small amplitudes we will observe that the system will be confined to a small neighborhood of the attractors. As we increase the noise level we will observe a critical value for which the system can escape from those attractors which are least robust against finite size stochastic perturbations. This allows a computational classification of the persistence of the attractors against repeated application of finite size shocks or, equivalently, the spread of uncertainty of future states of the system.

4. Using Machine Learning to Find Optimal Solutions

Human intuition and pattern recognition capabilities are very impressive in several different areas of problem solving: Skills are developed which allow us to solve recurrent problems (like playing video games) with amazing efficiency. We are also able to combine efficient solutions to seemingly unrelated problems and synthesize them to find new solutions to more complex problems. This is probably the way how most progress in science and technology is made. Finally, the human brain can come up with completely new and creative solutions which have practically nothing to with previous experiences and earlier solutions. These are exceptional cases which typically lead to break-through discoveries (while riding in the subway or looking at clouds)[7]. Limitations become evident, however, when the factors, to consider for the problem solution, exceed a certain critical complexity. Then we often regress to personal experiences, folk wisdom, astrology etc. On the other hand there exists a long tradition in creating "problem solving tools" which primarily enhance our information storage capacity [8]: Starting from rhymes and songs (utilizing the associative capacities of our memory) to more rudimentary and non-elegant tools like pencil and paper, tools which were looked down upon by representatives of "pure thought" schools, probably similarly to the situation today, where computers are often not considered to provide results with a similar level of mathematical rigor than results obtained solely with pencil and paper.

Today there are methods available which simulate on computers some aspects of the first three categories of human problem solving. A very widely used method in Artificial Intelligence explicitly borrows from what we know about brain functions related to human learning: Neural nets have found applications in many diverse fields including some of the AI features of the smart weapons used in the recent gulf war (see [11] for a recent overview). It is in this context interesting to note that apparently chaotic solutions in the dynamics of a neural network can be utilized for acceleration of pattern learning [33].

We have used a different approach, based on genetic algorithms, for searching in a 15-dimensional space of parameters and initial conditions of the generalized Richardson equations of three nations [12] for solutions which minimize imbalances in armament levels.

In such cases, the parameter space is so large that it is infeasible to search it exhaustively. Since each equation must be iterated until it reaches asymptotic behavior for many different initial conditions and tested for sensitivity (by seeing whether small changes of initial conditions lead to qualitatively different asymptotic behavior), each point in the parameter space is expensive to evaluate. Genetic algorithms offer the possibility of searching the parameter space intelligently to find regions of interest. As an estimate, we iterate the model for twenty time steps and compare the spending of X (the dominant country) with that of Y and Z (the allied countries). The imbalance function that we want to minimize is given by

$$F = |x_{20} - (y_{20} + z_{20})|.$$

---

[7]There is some evidence that certain chaotic dynamics features are used by the brain itself in order to make best "sense of the world" (see e.g. [14], [26])

[8]There is a rule that we can only spontaneously grasp and remember information which does not exceed the equivalent of 7 decimal digits, like in US phone numbers.

In the model there are twelve independent parameters (fifteen if the initial conditions are included). These real-valued parameters are defined to be in the range $(0, 1]$, so it is straightforward to discretize them into bins. This results in one binary-coded integer for each parameter designating its bin. The size of the bin is determined by the number of bits used in the discretisation (for example 8 bits corresponding to $2^8 = 256$ different bins for each parameter were used in [12]). The genetic algorithm finds balance-of-power solutions which were not known to exist previously. Manually setting the parameters had not suggested that such points existed. These results are non-trivial in that the genetic algorithm found many different solutions and most of the solutions were not degenerate (that is, parameters were not set to 0.0 or 1.0). This is in contrast to a similar search using neural networks, which has produced results which often suggest trivial solutions. Studying hyperplanes of parameter space shows that it contains regular subregions with close to perfect fitness. That suggests for our model that once we have found one solution with the genetic algorithm, it is easy to find other solutions in a neighborhood due to the smoothness of the system (see Fig. 9).

Closely connected with the development of chaos theory was the breath taking progress in the performance of computer hardware and software: Many results would have been practically impossible without the availability of fast, interactive computer graphics and scientific visualization tools. Currently there are developments are under way to utilize also audio senses to represent complex, multi dimensional data structures. Similar progress has been made in business areas where decisions are made with increasing reference to computerized data bases and electronic spreadsheets. As in many parallel developments we have to expect some convergence and synthesis: problems which require decisions with far reaching consequences need to be presented in a way that the decision maker can directly relate to them as in electronic spreadsheets. Traditionally these were static, did not include statistical analysis and dynamic simulations. Modern, spreadsheets, however, incorporate many features from simulation and visualization requirements. Thus we can formulate the Richardson arms race models in terms of a spreadsheet and thus are able to combine them in a natural way with existing methods of analysis and data handling (see e.g. Fig. 1).

5. Integration of Models Into Global Networks

The models and techniques discussed above only describe a very limited attempt to use recent developments of nonlinear dynamics research to describe complex problems of modern societies. It is clear that none of these aspects of international security can be seen in isolation. There allways have been very strong couplings between strategic questions and economical issues.

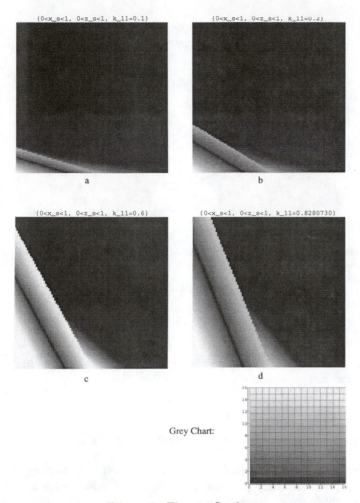

Figure 9: Fitness Surface

This figure illustrates the structure of the fitness function in a 2-dimensional hyper-plane through one of the points , found by the GA, for which the fitness function is zero within our resolution. That means 12 parameters are kept fixed at the GA values but parameters x_s, z_s are varied in the interval [0,1] whereas the value of k_{11} is kept fixed at (a) $k_{11} = 0.1$, (b) $k_{11} = 0.2$, (c) $k_{11} = 0.6$, (d) $k_{11} = 0.8$.

The axes represent the unit interval [0, 1] (evaluated at a resolution of $\delta x = \delta y = 0.01$). The fitness of each point in the plane is represented by its grey value: grey values go from black (0) to white (maximal value for the fitness parameter).

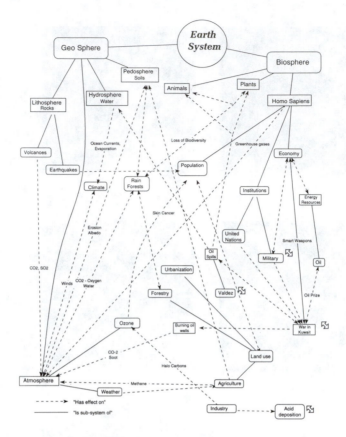

Figure 10: Diagram Interface for Model Network

Graphic display of diagram of different elements connected to international security problems. Each of the nodes with a hollow double arrow attached to it contains either data information (multi-media format) or launches (possibly on a remote machine) a simulation of a model of the area represented by the network-node.

In the future we can expect that changes in the global environment and population growth will enhance the stress on human living conditions and therefore become more tightly coupled to international, but also national security problems. The goal therefore has to be to find ways of linking different models, simulation and visualization tools as well as data bases and news services tightly together (Fig.10, see also e.g. [13], [32]). Fortunately it seems that modern technology in computation and communication make rapid progress towards a realization of the technical means to approach those goals (see e.g. [25]).

References

[1] R. Abraham, A. Keith, M. Koebbe, G. Mayer-Kress, Double Cusp Models, Public Opinion, and International Security, Intl. J. Bifurcations and Chaos (to appear 6/1991)

[2] S. J. Brams, "Superpower Games", Yale University Press 1985

[3] D.K. Campbell, "Nonlinear Science: from Paradigms to Practicalities", 218-262 in "From Cardinals to Chaos: Reflections on the Life and Legacy of Stanislaw Ulam", (Cambridge University Press 1989).

[4] D. Campbell, G. Mayer-Kress, "Chaos and Politics: Simulations of Nonlinear Dynamical Models of International Arms Races", Proceedings of the United Nations University Symposium "The Impact of Chaos on Science and Society", Tokyo, Japan, 15-17 April 1991

[5] G. Canavan, H. Flicker, O. Judd, K. Taggert, "Comparison of Analyses of Strategic Defense" Los Alamos National Laboratory Report P/AC-85-27, 1985

[6] G. Canavan, "Simple Discussion of the Stability of Strategic Defense", Los Alamos National Laboratory Report P/AC-85-81, 1985

[7] N. Choucri, T.W. Robinson (Eds.), "Forecasting in international relations", Freeman and Co., San Francisco 1978

[8] P. Collet, J.-P. Eckmann, "Iterated Maps on the Interval as Dynamical Systems", Birkhäuser, Boston, (1980)

[9] J.P. Crutchfield, J.D. Farmer, N.H. Packard, R.S. Shaw, "Chaos", Scientific American, 46-57, 1987

[10] J.-P. Eckmann, Roads to turbulence in dissipative dynamical systems, Rev. Mod. Phys. 52, 643-654, 1981

[11] J.D. Farmer, "A Rosetta Stone of Connectionism", Physica 42 D, 153-187, 1990

[12] S. Forrest, G. Mayer-Kress, Using Genetic Algorithms in Nonlinear Dynamical Systems and International Security Models, in: The Genetic Algorithms Handbook, L. Davis, (Ed.),Van Nostrand Reinhold, New York 1991

[13] J.A. Forrester, "World Dynamics", Wright-Allen Press, Cambridge Mass 1971

[14] W.J. Freeman, C.A. Skarda, "How the Brain Makes Chaos in Order to Make Sense of the World", Behavioral and Brain Sciences, 10, No. 2, 161-195, 1987

[15] C. Grebogi, E. Ott, J. Yorke, "Crises, Sudden Changes in Chaotic Attractors and Transient Chaos, Physica **7D**, 181, (1983)

[16] S. Grossmann, G. Mayer-Kress, "The Role of Chaotic Dynamics in Arms Race Models", Nature, **337**, 701-704, (1989)

[17] H.Haken, G. Mayer-Kress, Chapman-Kolmogorov Equation and Path Integrals for Discrete Chaos in the Presence of Noise, Z.Phys.B-Cond. Matter, 43, 183 (1981)

[18] M.D. Intriligator, D.L. Brito, "Can Arms Races Lead to the Outbreak of War?", J. of Conflict Resolution **28**, 63-83, 1984

[19] M.N. Kadyrov,"A Mathematical Model of the Relation between Two States, Global Development Processes ", Institute for System Studies, Preprint 1984

[20] J. Krause, G. Mayer-Kress, in preparation

[21] G. Mayer-Kress, H. Haken, "The influence of noise on the logistic system", J. Stat. Phys. **26**, 149, (1981)

[22] G. Mayer-Kress, H. Haken, "Attractors of convex maps with positive Schwarzian derivative in the presence of noise", Physica **10D**, 329-339, 1984

[23] G. Mayer-Kress, H. Haken, "An Explicit Construction of a Class of Suspensions and Autonomous Differential Equations for Diffeomorphisms in the Plane" Commun. Math. Phys. **111**, 63-74, 1987

[24] G. Mayer-Kress, "A Nonlinear Dynamical Systems Approach to International Security", in: "The Ubiquity of Chaos", S. Krasner (Ed.), Proceedings AAAS conference, San Francisco, January 1989, Los Alamos preprint LA-UR-89-1355

[25] G. Mayer-Kress, "EarthStation, an Installation at ARS electronica 'Out of Control', Linz, Austria, Spetember 10-13, 1991", informal distribution

[26] J. Nicolis, G. Mayer-Kress, G. Haubs, "Non-Uniform Chaotic Dynamics with Implications to Information Processing", Z. Naturforsch. 38a, 1157-1169, 1983

[27] L.F. Richardson, "Arms and insecurity", Boxwood, Pittsburgh 1960

[28] A. Saperstein, "Chaos - a Model for the Outbreak of War", Nature **309**, 303-305, 1984

[29] A. Saperstein, G. Mayer-Kress, "A nonlinear dynamical model of the impact of SDI on the arms race", J. Conflict Resolution, **32**, 636-670, 1988

[30] H. G. Schuster, Deterministic Chaos, An Introduction, Physik-Verlag: Weinheim, 1984.

[31] D. Sholl, G. Doolen, Y. C. Lee, "Analytic Solutions for Lanchester Equations"; Los Alamos Report, LAUR 90-959; Submitted to Journal of Operations Research

[32] T. Utsumi,J. DeVita, "GLOSAS Project." Computer Networks and Simulation II, ed. S. Schoemaker, 279-326. Amsterdam: North-Holland Publishing Co., 1982

[33] Paul Verschure, "Chaos based learning", Submitted to 'Complex Systems'

[34] K. Waltz, "Theory of International Politics", Addison-Wesley Publishing Company, Reading Mass. 1979

Part III
Economic Systems

Chaotic Dynamics in Economic Equilibrium Theory

M. Boldrin

Economic theorists have traditionally concentrated their attention on static phenomena or, at most, on dynamic behaviors that converge to stationary states. Recent literature has instead emphasized the possibility of persistent oscillations and chaos in dynamic economic models.

The major achievement of this research effort has been to show that persistent oscillations in economic variables are not incompatible with economic rationality and the working of the market mechanism.

The results presented are purely theoretical: abstract mathematical models describing various aspects of a competitive economy are studied in order to show that their solutions are dynamical system where nonlinearities are important.

Existing results in nonlinear dynamics are then invoked to claim the existence of cycles and chaos. No claim is made that such a class of models can be implemented in empirical applications.

1. INTRODUCTION

A proper assessment of the impact that chaotic dynamics has had on the field of economic theory requires an understanding of what the dominant paradigm of research in economic dynamics was, and still is today. In fact, the idea that market mechanisms are inherently dynamically unstable has played only a minor role in studies of economic fluctuations over the past quarter century. Instead, the dominant strategy, both in equilibrium business cycle theory and in econometric modeling of aggregate fluctuations, has been to assume model specifications for which the stationary equilibrium is determinate and intrinsically stable, so that in

M. Boldrin: Department of Managerial Economics and Decision Sciences, Kellogg Graduate School of Management, Northwestern University, Evanston, IL 60208.

the absence of continuing exogenous shocks the economy would tend toward a steady state growth path. The presence of a stationary pattern of fluctuations is then attributed to the existence of exogenous shocks of one kind or another—most often either technology or taste shocks, or stochastic shifts in government policies.

Recent work, however, has seen a revival of interest in the hypothesis that aggregate fluctuations and other exotic economic phenomena might have an endogenous character that would persist even in the absence of stochastic "shocks" to the economy. Even without giving credence to the extreme (and surely implausible) view that macroeconomic fluctuations can be interpreted as purely deterministic, the possibility that exogenous shocks might play a relatively minor role in generating the size of aggregate fluctuations that we observed must be judged of no small importance.

The endogenous cycle hypothesis is not totally new. Indeed, the earliest formal models of business cycles were largely of this type, including most notably the business cycle models proposed by Sir John Hicks, Nicholas Kaldor, and Richard Goodwin. In all these models the stationary growth path of the economy is locally unstable, but deviations from it are eventually contained by "floors and ceilings," such as shortages of productive factors on the upside or technological limits to the amount by which investment can be made negative on the downside.

By the late 1950s, however, this way of attempting to model aggregate fluctuations had largely fallen out of favor, the dominant approach having become instead the Slutsky-Frisch-Tingergen methodology of exogenous stochastic "impulses" that are transformed into a characteristic pattern of oscillations through the filtering properties of the economy's linear "propagation mechanism."

A major reason for the decline from favor of the endogenous cycle hypothesis concerns the inadequate behavioral foundations of the early models of this kind. The stability results obtained for many simple equilibrium models based upon optimizing behavior with perfect foresight —in particular the celebrated "Turnpike theorems" for optimal growth models doubtless led many economists to suppose that the endogenous cycle models were not only lacking in explicit foundations in terms of optimizing behavior, but also depended upon behavioral assumptions that were necessarily inconsistent with optimization. Dismantling this long-held prejudice has probably been the single most important achievement of the recent wave of studies in nonlinear economic dynamics. In fact, what the last ten to fifteen years of theoretical research has done is to pick up the standard, commonly-accepted models of intertemporal equilibrium and show that endogenous fluctuations and chaos were perfectly consistent with them. No area of economic theory and no class of models has been exempted by this revisionist enterprise and it seems now widely accepted, among economic theorists at least, that chaotic dynamic patterns <u>may</u> theoretically arise in a well functioning market economy.

The progress made, nevertheless, stops exactly here: still we have not shown that chaos does emerge in accepted economic models at empirically relevant parameter values and that, more importantly, the time series so obtained "match" (in any reasonable sense) the measured statistics of actual economics.

Our discussion will proceed in the following way. We begin by recognizing that, at least in the basic one-sector, neoclassical growth model it is true that cycle or chaotic policy functions cannot be optimal. Next we argue that this is indeed the only relevant economic framework in which such a critique holds. We show that any kind of dynamic behavior could be optimal in a growth model with two or more sectors of production. The global stability property of the one-sector model should therefore be considered as a very special implication of the highly restrictive hypotheses upon which it is built. Building on this general possibility theorem, we describe recent results on the presence of cycles and chaos in parameterized versions of economic models of intertemporal optimization.

2. THE ONE SECTOR MODEL

Day (1982) considers a one-sector, neoclassical growth model in which the dynamics of capital accumulation has the form:

$$k_{t+1} = s(k_t)f(k_t)/(1 + \lambda) = h(k_t) \tag{2.1}$$

where s is the saving function, f the production function, and $\lambda > 0$ is the exogenous population's growth rate. This is a discrete-time version of the famous Solow growth model. In the discrete-time form (2.1) Solow's assumption of a constant, exogenous saving rate and of a neoclassical, concave production function give rise to a map, $h(k_t)$, which is monotonically increasing and has one and only one interior steady state, $k^* = h(k^*)$. Therefore, not even damped oscillations are possible in this case.

The trouble with Solow's model is that it is not an optimizing one, i.e., the aggregate saving function is not explicitly derived from considerations of intertemporal efficiency. One is therefore free to pick "reasonable" shapes for $s(k_t)$ (and $f(k_t)$, obviously) in order to prove a claim. A typical Solow-like pair would be a constant savings ratio σ and a Cobb-Douglas form for f. (2.1) then becomes

$$k_{t+1} = \sigma B k_t^{\beta}/(1 + \lambda) \qquad (2.2)$$

which is monotonic and therefore stable. The first modification Day suggests is to the production function. By introducing a "pollution effect" in it one obtains:

$$k_{t+1} = \sigma b k_t^{\beta}(m - k_t)^{\gamma} / (1 - \lambda) \qquad (2.3)$$

which is unimodal and has period-three for certain ranges of parameter values. Returning to the Cobb-Douglas form and allowing instead for a variable saving rate, $s(k) = a(1 - b/r)k/y$, Day obtains:

$$k_{t+1} = a/(1 + \lambda) \; k_t [1 - (b/\beta B) k_t^{1-\beta}] \qquad (2.4)$$

using the fact that the rate of interest must be $r = \beta y/k$. This equation also displays chaos for feasible parameter values.

Day's examples (as well as many others) show that extremely simple behavioral hypotheses and model structures can produce very complicated dynamics. However, one may question whether the sort of behaviors assumed is in fact consistent with optimization within the given environment. For example, the assumption of a constant saving ratio was often used in the early "descriptive" growth models and can indeed be derived from intertemporal utility maximization under certain hypotheses (for example, logarithmic utility function) but it becomes especially implausible when a production function of the type embodied in (2.3) is proposed. Why should a utility-maximizing agent ever save up to the point at which marginal returns to capital are negative if he can obtain the same output level with much less capital stock? This implies that (given the assumptions on the technology) a "policy function" of the type (2.3) would never occur in an "optimal growth model" of the Cass (1965) type.

Although it is less obvious, Day's case of a variable saving ratio and a monotonic production function (i.e., Equation (2.4)) is equally inconsistent with intertemporal utility maximization. This was pointed out (in a general form) in Dechert (1984). Dechert's argument goes as follows: let $y_t = f(k_t)$ be total output at time t, as a function of the existing stock of capital. The consumer-producer chooses how to split it between consumption and future capital in order to maximize: $\sum_{t=0}^{\infty}$ $u(c_t)\delta^t$, where u is a concave utility function, δ is a time-discount factor, $\delta \in (0,1)$ and k_0 is given as an initial condition. It turn out that, even if the production function is not concave, the optimal program $\{k_0,k_1,k_2,\dots\}$ can be expressed by a policy function $k_{t+1} = \tau(k_t)$, which is monotonic. The dynamical systems induced in this way cannot therefore produce cycles or chaos. The economic prediction is that such a society will asymptotically converge to some stationary position. The latter is unique when f is concave.

3. THE MULTI-SECTOR MODEL

Here is the general model of intertemporal competitive equilibrium-optimal growth, of which the one sector model of Section 2 is just a special case (see Bewley (1982) or Boldrin-Montrucchio (1991, Ch. 1) for a more detailed illustration).

In every period $t = 0,1,2,\dots$ the representative agent derives satisfaction from a "consumption" vector, $c_t \in \mathbb{R}^m$, according to a utility function $u(c_t)$ which is taken increasing, concave and smooth as needed. The state of the world is fully described by a vector $x_t \in \mathbb{R}_+^n$ of stocks and by a feasible set $F \subset \mathbb{R}_+^{2n} \times \mathbb{R}^m$ composed of all the triples of today's stocks, today's consumptions, and tomorrow's stocks that are technologically compatible, i.e., a point in F has the form (x_t,c_t,x_{t+1}). Now define:

$$V(x,y) = \max_c u(c) \quad \text{s.t.} \quad (x,c,y) \in F$$

and let $D \subset \mathbb{R}_+^{2n}$ be the projection of F along the c's coordinates. Then V, which is the short-run or instantaneous return function, will give the maximum utility achievable at time t if the state is x and we have chosen to go into state y by tomorrow. It should be easy to see that to maximize the discounted sum $\sum_{t=0}^{\infty} u(c_t)\delta^t$ s.t. $(x_t,c_t,x_{t+1}) \in F$ is equivalent to max $\sum_{t=0}^{\infty} V(x_t,x_{t+1})\delta^t$ s.t. $(x_t,x_{t+1}) \in D$.

The parameter δ indicates the rate at which future utilities are discounted from today's standpoint (impatience): it takes values in $[0,1)$.

The following assumptions on V and D may be derived from basic economic hypotheses on u and F:

(A.1)   V: $D \to \mathbb{R}$ is strictly concave and smooth (if needed). $V(x,y)$ is increasing in x and decreasing in y.

(A.2) $D \subset X \times X \subset \mathbb{R}^{2n}_+$ is convex, compact and with nonempty interior. X
is also convex, compact and with nonempty interior.

The optimization problem we are facing can be equivalently described
as one of dynamic programming:

$$W(x) = \text{Max}\{V(x,y) + \delta W(y), \text{ s.t. } (x,y) \in D\}. \tag{3.1}$$

A solution to (3.1) will be a map τ_δ: $X \to X$ describing the optimal
sequence of states $\{x_0, x_1, x_2, \ldots\}$ as a dynamical system $x_{t+1} = \tau_\delta(x_t)$ on
X. The time evolution described by τ_δ contains all the relevant
information about the dynamic behavior of our model economy.

The question that concerns us is: What are the predictions of the
theory about the asymptotic behavior of the dynamical system τ_δ? Where
should a stationary economy converge under competitive equilibrium and
perfect foresight? A first, remarkable answer is given by the following
(see McKenzie (1986, 1987) for details and attributions):

Turnpike Theorem (discrete time): Under Assumptions (A.1) and (A.2) plus
smoothness of V there exists a value $\bar{\delta} < 1$ of the discount factor such
that for all the δ's in the interval $[\bar{\delta}, 1)$ the function τ_δ that solves
(3.1) has a unique, globally attractive fixed point $x^* = \tau_\delta(x^*)$.

But the turnpike property is not the end of the story. Many more
complicated dynamic patterns can be originated by τ_δ, for a fixed V and D,
as δ moves below $\bar{\delta}$. We will use the two-sector model as an illustrative
device. For a detailed analysis of the more general case, see Boldrin-
Montrucchio (1986), and Montrucchio (1990) for the continuous time case.

4. A TWO-SECTOR CHAOTIC ECONOMY

Here is a brief description of a two-sector economy. For more
details, see Boldrin (1986) and Boldrin-Deneckere (1990).

Two goods (a capital good and a consumption good) and only one
representative agent exist. The triples (w_t, r_t, q_t), $t = 0,1,2,\ldots$ denote
the labor wage rate, the gross capital rental and the price of capital in
every period t. They are expressed in units of the consumption good which
has the price fixed at one in all periods. We assume perfect foresight.

In each period the consumer is endowed with one unit of labor time,
which he supplies inelastically at the current wage rate, and with an
amount k_t of capital stock which is left over from pervious consumption-
saving decisions and that he supplies inelastically to the productive

sectors. His budget constraint is then:

$$c_t + q_t[k_{t+1} - \mu k_t] = r_t k_t + w_t$$

where $(1 - \mu)$ is the capital depreciation rate. Given the initial capital
stock k_0, the problem of the consumer amounts to pick up sequences of
consumption $\{c_t\}$ and gross saving $\{k_{t+1} - \mu k_t\}$ to maximize the present
value of his lifetime consumption stream under the period-by-period budget
constraint.

Goods are produced by two industries. We summarize this with two
production functions:

$$y^1 = F^1(k^1, \ell^1), \quad y^2 = F^2(k^2, \ell^2)$$

where the superscript 1 denotes the consumption sector and 2 denotes the
capital good sector; k^i, ℓ^i, $i = 1, 2$, are the quantities of capital and
labor used as inputs in either one of the two industries.

Firms take the price sequence $\{w_t, r_t, q_t\}$ as given. Their optimal
decision problems reduce to the choice of factors-demand sequences $\{k_t^i, \ell_t^i\}$
which maximize the present discounted value of the stream of future
profits. The economy's production possibility function is obtained as:

$$T(k_t, y_t) = \text{Max } F^1(\ell^1, k^1) \tag{T}$$
$$\text{s.t. } F^2(\ell^2, k^2) \geq y$$
$$\ell^1 + \ell^2 \leq 1, \quad k^1 + k^2 \leq x$$
$$\ell^1, k^1, \ell^2, k^2 \geq 0$$

The competitive equilibrium sequences $\{c_t, y_t, q_t, r_t, w_t, \ell_t^1, \ell_t^2, k_t^1, k_t^2\}_{t=0}^{\infty}$
may then be derived from the sequence of optimal capital stocks $\{k_t\}_{t=0}^{\infty}$
that solve:

$$W_\delta(k_0) = \text{Max } \Sigma_{t=0}^{\infty} V(k_t, k_{t+1}) \delta^t$$

$$\text{s.t. } (k_t, k_{t+1}) \in D$$
$$V(k_t, k_{t+1}) = T(k_t, k_{t+1} - \mu k_t)$$

using the following relations which hold either by definition or as a
condition for equilibrium:

$$c_t = V(k_t, k_{t+1}) \tag{4.1a}$$

$$y_t = k_{t+1} - \mu k_t \qquad (4.1b)$$

$$q_t = \delta W_\delta'(k_{t+1}) = -V_2(k_t, k_{t+1}) \qquad (4.1c)$$

$$r_t = V_1(k_t, k_{t+1}) \qquad (4.1d)$$

$$w_t = V(k_t, k_{t+1}) + q_t(k_{t+1} - \mu k_t) - r_t k_t \qquad (4.1e)$$

$$\ell_t^i = \ell^i(k_t, k_{t+1} - \mu k_t), \quad i = 1,2 \qquad (4.1f)$$

$$k_t^i = k^i(k_t, k_{t+1} - \mu k_t), \quad i = 1,2. \qquad (4.1g)$$

Let $\tau_\delta(\bullet): [0,\bar{k}] \to [0,\bar{k}]$ be the policy function associated with $W_\delta(\bullet)$. We will show that, given any C^2 function $\theta: X \to X$ with X a closed interval of the real line, one can find a two-sector economy and a discount factor $0 < \delta < 1$ for which such a θ is the optimal policy function τ_δ. This result is an application to this specific case of a general theorem first proved in Boldrin-Montrucchio (1986).

Proposition 1: Let the feasible set $D \subset X \times X$ be given, with X a closed interval of $\mathbb{R}$. Let $\theta \in C^2(X;X)$ be such that graph(θ) $\subset D$. Then there exists a short run return function $V: D \to \mathbb{R}$, continuous and strictly concave and a discount factor $\delta^* \in (0,1)$ such that $\theta = \tau_{\delta*}$, where $\tau_{\delta*}$ is the optimal policy associated to the given (D,V,δ^*). Moreover, $V(x,y)$ is increasing in x and decreasing in y, and one may recover from it the two production functions F^1 and F^2 satisfying (T1) and (T2).

Proof (Sketch): Let $\theta: X \to X$ be given. Consider the function:

$$U(x,y) = -(1/2)(y - \theta(x))^2 - (L/2)x^2 + mx$$

with L and m two positive real numbers. For L > 0 and large enough U is (strictly) concave in x and y. Also:

$$\text{Max}_y \ U(x,y) = U(x,\theta(x)) = W(x).$$

Let W so defined by the value function for the associated problem. A simple manipulation of equation (3.2) shows that the short-run return function $V(x,y) = U(x,y) - \delta W(y)$ in this case turns out to be:

$$V(x,y) = -(1/2)(y - \theta(x))^2 - (L/2)x^2 + mx(\delta L/2)y^2 - \delta my.$$

One needs to show that V is (strictly) concave and monotone in x and y for appropriately chosen L, m and δ. It is not very difficult to see that this is always possible and that the following estimate for δ^* and the associated L* are appropriate:

$$L^* = \eta\beta + \alpha\sqrt{\eta\beta}$$

$$\delta^* < [2\alpha\sqrt{\eta\beta} + \eta\beta + \alpha^2]^{-1}$$

where: $\eta = \max_D |y - \theta(x)|$, $\alpha = \max_X |\theta'(x)|$, and $\beta = \max_X |\theta''(x)|$.

Now, in order to recover the "fictitious" production functions F^1 and F^2 that give rise to the return function V, one may proceed in the following way. Set the depreciation rate for capital $(1 - \mu) = 1$ so that $k_{t+1} = y$ = output of $F^2(\ell^2, k^2)$. Then pick F^2 exactly as in our example, i.e., $y = \min\{1 - \ell^1, (x - k^1)/\gamma\}$. with γ a parameter in (0,1) to be defined later. Such a choice for F^2 obviously satisfies (T1) and (T2). In order to recover $F^1(\ell^1, k^1)$ one has to repeatedly substitute for $y = 1 - \ell^1$ and $x = \gamma y + k^1 = \gamma(1 - \ell^1) + k^1$ in the definition of V(x,y), therefore obtaining:

$$F^1(\ell^1, k^1) = -(1/2)[(1 - \ell^1) - \theta(\gamma - \gamma\ell^1 + k^1)]^2 \qquad (4.2)$$
$$+ m(\gamma - \gamma\ell^1 + k^1) - (L^*/2)(\gamma - \gamma\ell^1 + k^1)^2 + (\delta^*L^*/2)(1 - \ell^1)$$
$$- \delta^*m(1 - \ell^1).$$

Some tedious algebra will show that, given δ^* and L^* as defined earlier one can always pick $\gamma(\delta^*, L^*) \in (0,1)$ and $m(\delta^*, L^*) > 0$ such that F^1 as defined in (4.2) also satisfies (H1) and (H2).   Q.E.D.

I will now proceed to the study of a stylized example.

Example: Let $V(k_t, k_{t+1})$ be given by:

$$[a(k_t - \gamma k_{t+1})^\rho + (1 - a)(1 - k_{t+1})^\rho]^{1/\rho}.$$

Simple manipulations of the Euler equation:

$$-[a(k_{t-1} - \gamma k_t)^\rho + (1 - a)(1 - k_t)^\rho]^{(1/\rho)-1} \times \qquad (4.3)$$
$$\{(1 - a)(1 - k_t)^{\rho-1} + a\gamma(k_{t-1} - \gamma k_t)^{\rho-1}\} +$$
$$\delta a[a(k_t - \gamma k_{t+1})^\rho + (1 - a)(1 - k_{t+1})^\rho]^{(1/\rho)-1}(k_t - \gamma k_{t-1})^{\rho-1} = 0$$

will show that the unique, interior steady state $k(\delta)$ can be expressed as

$$k(\delta) = \{1 + (1 - \gamma)[(1 - a)/a(\delta - \gamma)]^{1/(1-\rho)}\}^{-1}. \qquad (4.4)$$

Therefore, for $\delta \in [0,\gamma]$ we have no interior steady state and for $\delta \in (\gamma,1)$ we have a unique interior steady state, which is on the upward sloping branch of τ_δ for $\delta < \gamma(1 + (1 - a)/a\gamma^\rho)$ and on the downward sloping one for $\delta > \gamma(1 + (1 - a)/a\gamma^\rho)$. Let us simplify the algebra by setting $\rho = 0$ in the definition of F^1. This is the so-called Cobb-Douglas case. The function $V(x,y)$ is now: $(1 - y)^\alpha(x - \gamma y)^{1-\alpha}$ with $\alpha = 1 - a$.

The proofs of the following list of results may be found in Boldrin-Deneckere (1990).

<u>Proposition 2</u>: For $0 < \delta \leq \gamma$ the optimal path k_t converges to zero for any initial condition k_0 in $[0,1]$, and no interior OSS exists. For $\gamma < \delta \leq \gamma/(1 - \alpha)$, there exists a unique interior OSS k^* and the optimal path converges to k^*, for any k_0 in $(0,1]$.

Let us now turn to the case where $\delta > \gamma/(1 - \alpha)$. It is well known from the literature on optimal growth theory that when k^* is locally stable for τ_δ, i.e., $\partial\tau_\delta(k)/\partial k_{k=k^*} < 1$, the second order system produced by the Euler equation has a local saddle point structure at k^*. This means that of the two eigenvalues of the characteristic polynomial:

$$\delta V_{12}(k^*,k^*)\lambda^2 + [V_{22}(k^*,k^*) + \delta V_{11}(k^*.k^*)]\lambda + V_{12}(k^*,k^*) = 0 \qquad (4.5)$$

associated with the linearization of (4.3), one lies inside, and one lies outside the unit circle. In fact, the smaller eigenvalue corresponds to $\tau_\delta'(k^*)$ when the latter exists. What happens when $\delta \in [\gamma/(1 - 2\alpha),$ $(a + \gamma)(1 - \alpha)]$? A partial answer is the following proposition.

<u>Proposition 3</u>. Let $\alpha < (1 - \gamma)/2$. Then the policy function τ_δ has a cycle of period two for δ in a neighborhood of $\delta^- = \gamma/(1 - 2\alpha)$ and $\delta^+ = (\alpha + \gamma)/(1 - \alpha)$. These cycles are locally stable when they exist for $\delta \in (\delta^-,\delta^+)$, and unstable in the other cases.

The reader should note that, for given $\delta < 1$, it is always possible to choose α and γ in $(0, (1 - \gamma)/2)$ and in $(0,1)$, respectively, such that $\gamma/(1 - 2\alpha) = \delta$. This means that at every level of discounting we can always find <u>some</u> technology that has optimal cycles! In fact, the dynamic behavior of this economy for $\delta \in (\delta^-,\delta^+)$ may become very complicated. Our contention is that, for suitable α and γ, there exists an interval $(\delta^*,\delta^{**}) \subset (\delta^-,\delta^+)$ at which τ_δ has (at least) topological chaos. We demonstrate this at the end of this section. We also believe that the emergence of chaos follows the classical "period-doubling bifurcation pattern" as $\delta \to \delta^*$ from the left or $\delta \to \delta^{**}$ from the right. Evidence from the simulations we have run supports this contention. Here we only show that a second flip bifurcation may lead to an orbit of period four.

Proposition 4: Let $x(\delta), y(\delta)$ denote an interior period two orbit of τ_δ, for δ values in (δ^-, δ^+). Let $V_{ij}^* = V_{ij}(x(\delta), y(\delta))$, and $V_{ij} = V_{ij}(y(\delta), x(\delta))$, $i, j = 1, 2$. Assume there exists an interval $[\delta^{--}, \delta^{++}] \subset (\delta^-, \delta^+)$ and a $\delta^0 \in (\delta^{--}, \delta^{++})$ such that the function $G(\delta) = V_{22}^* V_{22} + \delta^2 V_{11}^* V_{11} + \delta(V_{11}^* V_{22}^* - V_{12}^{*2}) + \delta(V_{11} V_{22} - V_{12}^2) + (1 + \delta^2)V_{12}^* V_{12}$, satisfies:

$$
G(\delta) \begin{cases} > 0 & \text{for} \quad \delta \in [\delta^{--}, \delta^0) \\ = 0 & \text{for} \quad \delta = \delta^0 \\ < 0 & \text{for} \quad \delta \in (\delta^0, \delta^{++}) \end{cases}
$$

Then there exists a period four orbit for τ_δ bifurcating from $(x(\delta), y(\delta))$ at $\delta = \delta^0$.

To verify the presence of such bifurcations in our model, consider the example $\alpha = .03$, $\gamma = .09$. Proposition 2 implies that the steady state k^* is locally stable when δ lies in the interval $[.0928, .0957]$. For discount factors in $[.0957, .0974]$ stable period 2 orbits are present, verifying Proposition 3. At $\delta = .0974$, the period two orbit $x^* = .0738$, $y^* = .3980$ bifurcates into a stable period four orbit, which exists for $\delta \in [.0974, .0978]$. In fact, our simulation results reveal that successive bifurcations eventually lead to chaos when δ reaches the value .099. This chaos exists for $\delta \in [.099, .112]$, as can be checked directly. As expected, the concavity of V implies that only extreme values of the parameters can yield chaotic dynamics.

One might suspect that part of the reason that such extreme values of the parameters are needed stems from the fact that the elasticity of substitution between capital and labor in the consumption good sector is fairly large. To investigate this issue, we also ran simulations for a tractable generalization of our model, which retains the Leontief technology for the investment sector, but allows for a CES in the consumption sector. Thus, $F^1(\ell^1, k^1) = [\alpha(\ell^1)^\rho + (1 - \alpha)(k^1)^\rho]^{1/\rho}$. The elasticity of substitution, σ, for the CES is equal to $1/(1 - \rho)$; negative values of ρ thus permit much smaller values of σ. The simulations revealed that chaotic optimal paths do arise for this model as well, and that when σ is fairly low, chaos appears for values of the discount factor roughly three times larger than those found for the Cobb-Douglas model. A typical example has $\alpha = .2$, $\gamma = .2$, $\delta = .25$, and $\rho = -.5$. Since the values for the discount factor at which chaos appears in the Cobb-Douglas model are themselves approximately 100 times larger than the ones found in the artificial economies constructed by Boldrin and Montrucchio (1986) and Deneckere and Pelikan (1986), no definite conclusion can be drawn, at this stage, as to whether a model of this type could produce chaotic dynamics at more reasonable values of the discount factor.

5. CONCLUSION

In this paper, I illustrated a simple general equilibrium model which produces unique, but sometimes cyclical and chaotic paths for aggregate variables such as output, consumption, and investment. Despite the fact that an analytical expression for the policy function is unavailable, we were able to characterize the dynamic behavior of our economy in terms of its basic parameters: α, the labor share of income in the consumption sector; γ, the capital/labor ratio in the investment sector; and δ, the discount factor. For many values of the parameters, the unique steady state was shown to be globally asymptotically stable. For other values of the parameters, we obtained a unique period two point, which was globally attractive. Successive bifurcations then led to a chaotic regime, but only for rather unrealistic values of the parameters. This fact casts some doubt on the notion that, in one-dimensional capital good models, chaos is a useful way to model the apparently self-sustained nature of the trade cycle. The highly nonlinear bell-shaped form for the policy function that is necessary in order to produce complex dynamics forces one to resort to rather unrealistic values of the parameters. Future research on models with a higher dimensional state space may be more successful in this regard.

While still in its infancy, the study of nonlinearities in economic models is likely to provide insights into the forces behind observed economic fluctuations. In our model, we underlined the importance of intersectoral substitution effects (induced by different degrees of profitability in different sectors) as well as intertemporal substitution effects in determining factor allocation decisions, investment activities, and so on.

References
1. Benhabib, Jess and Kazuo Nishimura, 1985, "Competitive Equilibrium

Cycles," Journal of Economic Theory, 35, 284-306.

2. Boldrin, Michele, 1989, "Paths of Optimal Accumulation in Two-Sector
   Models," in: William Barnett, John Geweke and Karl Shell, eds.,
   Economic Complexity: Chaos, Sunspots, Bubbles and Nonlinearity
   (Cambridge University Press, Cambridge).

3. Boldrin, Michele and Luigi Montrucchio, 1986, "On the Indeterminacy of
   Capital Accumulation Paths," Journal of Economic Theory, 40, 26-
   39.

4. Boldrin, Michele and Raymond Deneckere, 1990, "Sources of Complex
   Dynamics in Two Sector Growth Models," Journal of Economic
   Dynamics and Control, 14, 627-654.

5. Day, R., 1982, "Irregular Growth Cycles," American Economic Review,
   72, 406-414.

6. Deneckere, Raymond and Steve Pelikan, 1986, "Competitive Chaos,"
   Journal of Economic Theory 40, 13-25.

7. Goodwin, Richard M., 1982, Essays in Economic Dynamics (McMillan,
   London).

8. McKenzie, Lionel W., 1986, "Optimal Economic Growth, Turnpike Theorems
   and Comparative Dynamics," in: Kenneth Arrow and Michael
   Intriligator, eds., Handbook of Mathematical Economics, Vol. III,
   (North Holland, Amsterdam).

9. Montrucchio, Luigi, 1990, "Dynamical Systems that Solve Continuous
   Time Concave Optimization Problems: Anything Goes," mimeo,
   University of Torino, Italy.

Chaos and the Foreign Exchange Market

C. Larsen and L. Lam

In the past, complex phenomena have often been assumed to be the result of stochastic processes. Because systems that exhibit chaos appear stochastic, measures are needed to differentiate among them. A common measure, the correlation dimension, is determined from the time series generated by daily foreign exchange spot prices for the Swiss Franc, Japanese Yen, and French Franc. The correlation dimension is found to saturate at $d_c \approx 2.8$ for the Swiss Franc and Japanese Yen, and at $d_c \approx 2.1$ for the French Franc.

1. INTRODUCTION

The realization within the last decade that simple deterministic nonlinear systems can exhibit behavior which looks stochastic in nature has motivated the development of tests to discern between chaotic and stochastic dynamical systems. Often this determination must be made from the time series generated from the measurement of few or even a single variable with no prior knowledge of the inherent dynamics of the system. Hence, the correlation dimension introduced by GRASSBERGER and PROCACCIA [1] and put on a firm mathematical foundation by TAKENS [2] has become a popular test for chaos when knowledge of the dynamical system is limited to an experimental time series.

C. Larsen and L. Lam: Nonlinear Physics Group, Department of Physics, San Jose State University, San Jose CA 95192-0106.

2. CORRELATION DIMENSION

To extract information from a time series it is necessary to begin by geometrically reconstructing the attractor on which the trajectory producing the data is assumed to lie. This can be done by constructing n-dimensional vectors from the original time series. Starting with the original time series of measurements taken at regular time intervals, $\{m(1), m(2), ..., m(N)\}$; construct vectors $X(i) = [m(in+1), m(in+2), ..., m(in+n)]$, $i = 0,..., P-1$. In this manner the $X(i)$ are points to be associated with an n-dimensional projection of the phase space of the original dynamical system. Using the sequence $X(0), X(1), ..., X(P-1)$; define the functions C_i^n and C^n as follows:

$$C_i^n(r) = P^{-1}\{ \text{ number of } X(j) \text{ such that } d[X(i),X(j)] \le r \}, \qquad (1)$$

$$C^n(r) = P^{-1} \sum_i C_i^n(r)$$

$$= P^{-2}\{ \text{ number of ordered pairs } [X(i),X(j)]$$

$$\text{such that } d[X(i),X(j)] \le r \} \qquad (2)$$

where $d[X(i),X(j)]$ is any norm and can be interpreted as the distance between points $X(i)$ and $X(j)$ (Euclidean norm). The correlation dimension d_c is then defined as:

$$d_c = \lim_{r \to 0} \lim_{P \to \infty} \frac{\log C^n(r)}{\log r}. \qquad (3)$$

In practice, (3) must be modified as the limits are not approachable. Because P is finite, for small r there will be poor statistics setting a lower bound r_0 on the useful values of r. A very informative article on the use of small data sets is RAMSEY and YUAN [3]. An upper bound r_1 exists due to the existence of "nonlinear" effects. In the region (r_0,r_1) the correlation dimension is constant and can be found experimentally from the slope of log $C^n(r)$ vs. log r. A useful discussion of the above can be found in ECKMANN and RUELLE [4].

If the original time series is generated by an attractor of finite dimension, then the correlation dimension will saturate for some greater value. This can be easily understood since for sufficiently large n, the attractor is embedded in a space of greater dimension than itself. For purely stochastic systems the correlation dimension does not saturate and is equal to n.

3. NUMERICAL EXPERIMENT

Daily Interbank spot prices for the Swiss Franc, Japanese Yen, and French Franc in the years 1981 to 1983 [5] were used as the time series. Since the data was taken at regular intervals each day, vectors $X(i)$ had subsequent daily prices as components. The choice of the range (r_0, r_1) experimentally is problematic. The difficulty lies in the interpretation of for what values of r is the curve log $C^n(r)$ vs. log r linear. Also, as n is increased $d[X(i), X(j)]$ increases as well so the values of r for which the curve is linear increase in magnitude. To avoid the subjectivity of "eyeballing" the curve, the values of r were chosen so that $0.01 \leq C^n(r) \leq 0.1$. This linear region for the Swiss Franc in plotted in Fig. 1 for increasing n. Error associated with the choice of the range (r_0, r_1) is discussed in MAYER-KRESS [6].

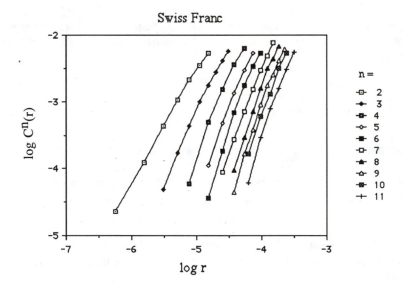

Fig. 1. Linear region of log $C^n(r)$ vs. log r for Swiss Franc with n varying between 2 and 11.

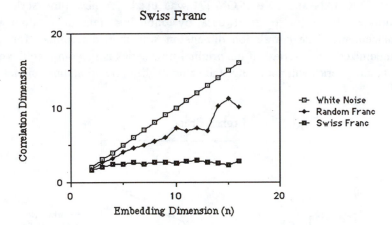

Fig. 2. Correlation dimension d_c vs. n for Swiss Franc. Also plotted are the correlation dimension of the Swiss Franc data which was randomized, and a purely random Gaussian White Noise for comparison.

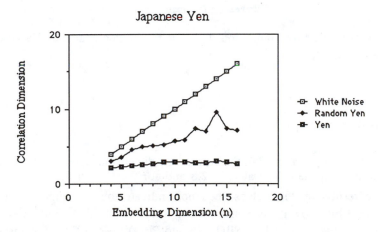

Fig. 3. Correlation dimension d_c vs. n for Japanese Yen. Also plotted are the correlation dimension of the Japanese Yen data which was randomized, and a Gaussian White Noise for comparison.

The correlation dimension was determined for increasing values of n by finding the slope of the curves in Fig. 1. and for the other currencies as well. The results are plotted in Figs. 2-4. For comparison, a method mentioned in

SCHEINKMAN and LEBARON [7] was used. A new time series was generated from the original data by choosing data at random with replacement. The correlation dimension was then determined for these "randomized" time series. If the original time series is random then the order of the time series will not change the value of the correlation dimension.

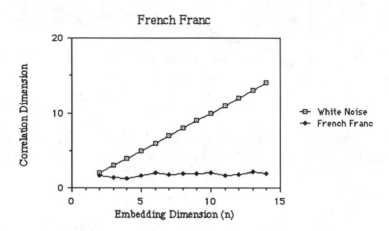

Fig. 4. Correlation dimension d_c vs. n for French Franc, and a Gaussian White Noise for comparison.

4. DISCUSSION

As can be seen in Figs. 2-4, the correlation dimension saturates for the Swiss Franc and Japanese Yen at $d_c \approx 2.8$ and for the French Franc at $d_c \approx 2.1$. These results by themselves need not indicate that the time series were produced by a low dimensional chaotic attractor. In fact, finite correlation dimensions were found by OSBORNE and PROVENZALE [8] for time series generated by stochastic processes that had power law type spectra; or so called "colored" noise. The temporal order of a time series generated by a "colored" noise would not change the value of the correlation dimension. However, the correlation dimensions for the randomized time series in Figs. 2-4 are significantly higher than the original series. Hence, definite temporal correlations do exist, consistent with data generated by chaotic dynamical systems. Further evidence of nonlinear correlations in daily foreign exchange rates can also be found in HSIEH [9].

While this does not prove that the Swiss Franc, Japanese Yen, and French Franc foreign exchange rates are purely chaotic dynamical systems, it does show that they are not purely stochastic and are hence deterministic in some manner. Evidence suggests that this determinism is of low dimensionality and plays a large part in determining the dynamics of foreign exchange rates. Whether or not models with genuine predictive power can be formulated remains to be seen.

Work supported by a Syntex Corporation Grant of Research Corporation.

REFERENCES

1. P. Grassberger and I. Procaccia, Physica **9D**, 189 (1983).
2. F. Takens, in *Dynamical Systems and Turbulence*, eds. D. Rand and L. S. Young (Springer, Berlin, 1981).
3. J. B. Ramsey and J. Yuan, Nonlinearity **3**, 155 (1990).
4. J. -P. Eckmann and D. Ruelle, Rev. Mod. Phys. **57**, 617 (1985)
5. *International Monetary Market Yearbooks*, Chicago Commodity Exchange Press, 1981-1983.
6. G. Mayer-Kress, in *Directions in Chaos,* vol.1, ed. Hao Bai-Lin (World Scientific, Singapore, 1987).
7. J. A. Scheinkman, and B. LeBaron, Journal of Business **62** (3), 311 (1989).
8. A. R. Osborne, and A. Provenzale, Physica **35D**, 357 (1989).
9. D. Hsieh, Journal of Business **62** (3), 339 (1989).

Part IV
Earthquakes and Sandpiles

Earthquakes as a Complex Phenomenon

C. Tang

Earthquakes are a very rich and complex phenomenon. It is essential to know what physical mechanisms are important. Modeling earthquake dynamics provide insights into the underlying physics. Some recent work will be reviewed.

1. INTRODUCTION

Earthquakes result from the movement of tectonic plates. These plates consist of the earth crust and the upper mantle and form the hard cover of the earth (the lithosphere). They drift slowly, with average rates of a few centimeters per year; the motion seems not complex at all, at least within a time scale of million years. However, since the plates move relatively to each other, stress may build up. Most of these stress are released through some kind of dynamical instabilities – usually sudden slips along the plate boundaries and occasionally new fractures within the plates – generating earthquakes. Earthquake, which occurs more than million times a year and in which the scale of energy release spans more than ten orders of magnitude, is a rather complex phenomenon indeed. For example, we do not know where, when and with what magnitude an earthquake will occur. On the other hand, it has certain characteristic patterns and follows some simple statistical laws. One of such laws, and perhaps the most remarkable one, is the Gutenberg-Richter law [1]:

Chao Tang: Institute for Theoretical Physics, University of California, Santa Barbara, CA 93106. Address after October 1, 1991: NEC Research Institute, 4 Independence Way, Princeton, NJ 08540.

$$N \propto M_0^{-B} \ , \tag{1.1}$$

where $M_0 = \mu \overline{\Delta u} A$ (with μ being the shear modulus, $\overline{\Delta u}$ the mean slip distance, and A the rupture area) is the seismic moment in an earthquake event which is proportional to the energy release, N is the number of earthquakes with seismic moment larger than M_0, and $B \approx 2/3$. The power-law distribution (1.1) implies that there is no typical scale for earthquakes. The lack of an apparent scale is reminiscent of the systems undergoing the continuous phase transitions. Indeed, it has been suggested that spatially extended dynamical systems may organize themselves into a critical state, and that scaling laws in nature, like (1.1), may well be the manifestations of the self-organized criticality [2]. The earth's hard cover (the lithosphere) can be viewed as such a dynamical system. On one hand, it is persistently being driven, by the tectonic motion, toward a dynamic instability. On the other hand, the instability (earthquakes) brings the system back to a new stable state. Hence, the system is always on the verge of stability (or instability) − in another words, it is critical. The critical state is characterized by complex fluctuations (earthquakes) of very wide range in magnitude. From this point of view, the complexity in earthquakes is generated dynamically, i.e., it is the intrinsic property of the dynamical system.

An extensive literature on earthquake modeling exists [3]. In the following, I will briefly review two recent approaches which were motivated by the above point of view and are, in some sense, complementary to each other. The emphasis will be on the scaling behavior of earthquakes. The first approach is more appropriate to a fault system, while the second concentrates on a single fault. The models involved are all very simple, too simple in many respects. Nevertheless, as we will see, these simple models are capable of generating rather complex behaviors which can be compared with real earthquakes. Furthermore, they can be used to suggest questions that we might ask about real seismological data. This simplicity also enables us to see which features of the model are relevant to which physical phenomena in real systems. It therefore makes sense to use these models as a reasonable starting point.

2. MODELING THE FAULT SYSTEM

Generalizing the work of BAK and TANG [4], CHEN, BAK, and OBUKHOV

[5] proposed the following model which focuses on an entire earthquake region (a fault system). As sketched in Fig. 1, a d-dimensional medium is subjected to a slowly increasing external stress field. The continuum medium is represented by an array of springs and (massless) blocks. When the stress somewhere exceeds a threshold value (which must be eventually since the stress is ever increasing), the spring breaks and the stress on that site is released, while the medium undergoes a local shear deformation (rupture). This breaking causes a very anisotropic redistribution of elastic forces. (The effect is equivalent to that of a dipole field, hence falling off roughly as $1/r^d$ with the distance from the instability.) The redistribution process takes place essentially with the speed of sound which is much faster than the geological time scale involved in the building up of stress. In this model this process is instantaneous through the entire system. The redistribution may cause the stresses on other sites to exceed their threshold values, resulting in further ruptures and stress redistributions, and so on — a local instability may trigger a chain reaction.

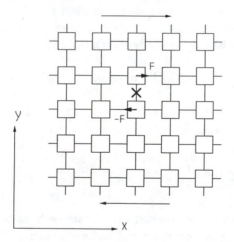

Fig. 1. Schematic illustration of fault region subjected to external shear stress. The vertical springs represent shear stress; the horizontal ones represent axial stress. The model is driven by slowly increasing the shear forces. When the local stress is larger than the threshold stress, the spring breaks, causing a redistribution of elastic forces. The broken bond heals immediately, i.e., it is replaced with a new spring of zero stress.

In order to calculate the force redistribution, consider a two-dimensional lattice (generalization to three-dimensional case is straightforward). The model is further simplified by considering only the component of the force in the direction of the external shear, say the x direction. Let $\sigma_x(r)$ be the (axial) stress in the bond between blocks at r and $r+e_x$ (e_x is the unit vector at x direction); this represent a force $\sigma_x(r)$ on the block at r and $-\sigma_x(r)$ on the block at $r+e_x$. Correspondingly, let $\sigma_y(r)$ be the (shear) stress in the bond between blocks at r and $r+e_y$. Then for the system at rest the following equations should be satisfied:

$$\sigma_x(r)+\sigma_y(r)-\sigma_x(r-e_x)-\sigma_y(r-e_y)=0, \qquad (2.1a)$$
$$\sigma_x(r)+\sigma_y(r+e_x)-\sigma_x(r+e_y)-\sigma_y(r)=0. \qquad (2.1b)$$

Eq. (2.1a) says that the total force on each block is zero. While (2.1b) is the statement that the sum of stress along a closed loop is zero (This can be achieved by rescaling x and y appropriately, since $\sigma \propto \Delta u$, where Δu is the difference between the displacements at two neighboring blocks (the strains)). This model does not describe the details of the dynamical process of ruptures, but only the stress distribution before and after each local rupture. Thus the effect of breaking a spring is approximated to be that of applying a pair of forces on the broken bond (See Fig. 1, where a vertical bond is broken). The additional stress on each site due to the breaking can then be calculated by the Green's function technique:

$$\sigma'_{x,y}(r)=-FG_{x,y}(r-r_0)+FG_{x,y}(r-(r_0+e_y)), \qquad (2.2)$$

where $G_{x,y}$ are the Green's functions of (2.1), which, in the limit of infinitely large medium, take the form

$$G_{x,y}(r)=\frac{1}{(2\pi)^2}\int_0^{2\pi}dk_x\int_0^{2\pi}dk_y\frac{(1-e^{ik_{x,y}})e^{i\mathbf{k}\cdot\mathbf{r}}}{4-2(\cos k_x+\cos k_y)}. \qquad (2.3)$$

The value of F can be determined self-consistently to be $F=-[d/(d-1)]\sigma_0$, where σ_0 is the stress on the bond before breaking. The broken bond heals immediately; it is replaced with a new spring of zero stress.

An earthquake event in this model is the chain reaction of bond breaking triggered by the breaking of one bond. The rate of the external shearing is made small enough so that no distinct events overlap in space and time. The threshold of each spring is a random number $\in [0,1]$. Starting from essentially any configuration, the system eventually enters into a statistically stationary state where the release of the stress due to "earthquakes" balances the increase of the stress due to the external driving. In the stationary state, the model exhibits a wide range of events: from single bond breaking to ruptures across the entire system.

Figure 2 shows a large (compared with the system size of the simulation) event. Note that the faults generated are not necessarily connected and appear to have certain structure. The total number of sites s which have ruptured in an event is used as a measure for the seismic moment or energy release. In Fig. 3 plotted is the differential distribution of s in a three-dimensional system. It seems to follow a power-law with a B-value 0.6, which is very close to (1.1). This agreement is rather remarkable, considering the simplicity of the model. Since the model completely ignored the details of dynamic rupture process, it suggests that the Gutenberg-Richter law (1.1) originates from the distribution and redistribution of the stress field, and from the interaction of faults. When

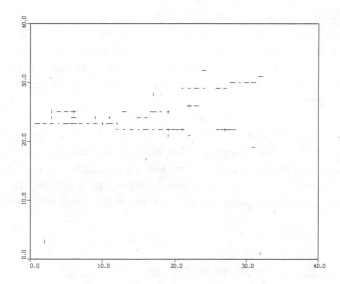

Fig. 2. The structure of a single earthquake. The line segments represent the displacement (proportional to the release of stress) at the sites that have ruptured.

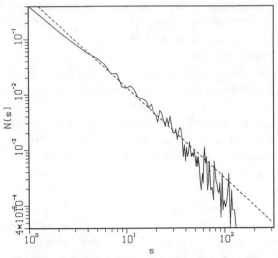

Fig. 3. Differential distribution of earthquakes generated in the stationary state of the model for a 20x20x20 lattice. The dashed line is a fit to a power-law with slope of -1.6. The fall-off at large sizes is due to the finite-size effect. The system appears to be in a critical state.

the broken bonds are replaced with weaker new springs so that subsequent earthquakes are likely to happen in some fault region which itself is generated dynamically, the scaling behavior remains the same. It may be sensible to use this model (or its modified versions which include asperity, creep, etc.) to study the formation, interaction, and evolution of a fault system.

The issue of the predictability of earthquakes has also been addressed within the context of this kind of models [5,6]. One question to ask is whether the system, once reached a critical state, is chaotic in the sense that it has a positive Lyaponov exponent. It is found that a small deviation in phase space grows not exponentially, but as a power-law in time.

3. THE BURRIDGE-KNOPOFF MODEL OF A SINGLE FAULT

The Burridge-Knopoff model [7] is a one-dimensional block and spring model for a single fault. Various versions of this model have been studied from various aspects by several authors [8]. Recently, CARLSON and LANGER [9], and later, CARLSON, LANGER, SHAW, and TANG [10] studied the scaling behavior of a uniform and deterministic version of the model. As illustrated in Fig. 4, the model consists of a chain of blocks of mass m coupled to each other by harmonic springs of strength k_c and attached to a fixed surface, shown above the blocks in the figure, by leaf springs of strength k_p. The blocks are in contact with a rough substrate which is moving at speed v to the left as shown. Equivalently, the substrate may be fixed and the blocks pulled to the right by the upper surface acting through the "pulling springs" k_p. In a dimensionless form, the equations of motion for this model are

$$\ddot{U}_i = l^2(U_{i+1} + U_{i-1} - 2U_i) - U_i - \phi(2\alpha v + 2\alpha \dot{U}_i),$$ (3.1)

where U_i is the displacement (relative to the fixed upper surface) of the ith block and $l^2 = k_c/k_p$. ϕ is the velocity-dependent "stick-slip" frictional force between the blocks and the substrate and is the key (and the only) nonlinear ingredient in the model. In the work to be described below, ϕ has the form

$$\phi(z) = \begin{cases} (-\infty, 1], & z = 0 \\ (1 - \sigma) / \{1 + [z/(1 - \sigma)]\}, & z > 0. \end{cases}$$ (3.2)

The static "sticking" friction $\phi(0)$ has a threshold value 1, while the dynamic "slipping" friction decays monotonically to zero from the initial value $\phi(0^+) = 1 - \sigma$. The "pulling" speed v that appears in (3.1) is the relative speed of the two tectonic plates and, accordingly, is of order 10^{-8} (in units of the slipping speed) or smaller. The quantity α in (3.1) is a measure of the "weakness" of the velocity-weakening friction: larger

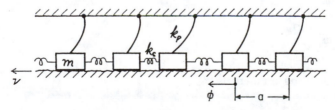

Fig. 4. Schematic illustration of the block and spring model of a single fault.

values of α mean weaker slipping friction and more pronounced instability. The "seismic moment" M of an event in this model is defined to be

$$M = a \sum_i \delta U_i , \qquad (3.3)$$

where the summation is over the blocks which have moved in the event. The corresponding "magnitude" $\mu = \ln M$.

It is helpful to think (3.1) as the discretized version of the partial-differential equation

$$\ddot{U} = \xi^2 \frac{\partial^2 U}{\partial s^2} - U - \phi(2\alpha v + 2\alpha \dot{U}), \qquad (3.4)$$

where $\xi = la$ is the sound speed (which has the dimension of length, since time is dimensionless in (3.1) and (3.4)). Hence, the parameter l in (3.1) is also the number of blocks in the length ξ, which must diverge like a^{-1} in the continuum limit. However, as we will see later, certain properties of (3.4) or (3.1) depend on the "ultraviolet cutoff" a.

One of the most dramatic features of this model is that it exhibits two qualitatively different kinds of earthquakelike events: relatively small, localized events; and large, delocalized events. The small events act mainly to smooth out certain region in preparation for a large event in which enough energy is released in the smoothed region to trigger a propagating rupture. A typical sequence of events is shown in Fig. 5. The area between consecutive curves in Fig. 5 is the moment M of the event which has caused the displacement. The crossover from "localized" to "delocalized" events can be estimated analytically. It happens roughly at events of size

$$\tilde{\xi} = \frac{2\xi}{\alpha} \ln\left[\frac{4l^2}{\sigma}\right]; \qquad (3.5)$$

the corresponding moment is

$$\tilde{M} \cong \frac{2\xi}{\alpha} . \qquad (3.6)$$

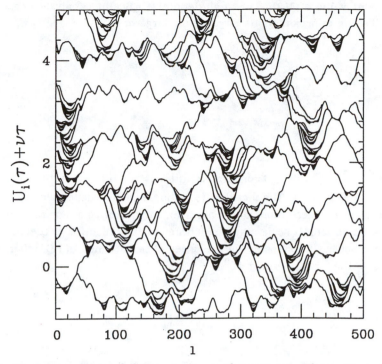

Fig. 5. A sequence of fully stuck configurations. There is a sharp distinction between the smaller, more abundant, localized events, which act mainly to smooth out the local minima in the configuration, and the larger delocalized events, in which the displacements are substantially larger. The results shown here are for $l=3$, $\alpha=1.1$, and $\sigma=0.01$.

Note the explicit l dependence of $\widetilde{\xi}$ in (3.5).

The frequency-magnitude distribution is plotted in Fig. 6, where $\mathscr{D}(\mu)d\mu$ is the number of events per unit fault length per unit displacement with magnitude in the interval $[\mu, \mu+d\mu]$. Several important features of the distribution can be noted. First, the minimum occurs around $\widetilde{\mu}=\ln \widetilde{M}$, indicating that localized and delocalized events have very different distributions. Second, the small, localized events, with $M<\widetilde{M}$, has a power-law distribution with $B\approx1$; the distribution of delocalized events, with $M>\widetilde{M}$, rise almost linearly and then are cut off sharply at a l-

dependent value of M that we shall denote M^*. Scaling analysis suggests the following approximations for $\mathcal{D}(\mu)$ for large α:

$$\mathcal{D}(\mu) \approx \begin{cases} \left(\dfrac{\tilde{M}}{M^*}\right)^2 \dfrac{2}{M} \propto \dfrac{1}{l^2 M} \,, & M_1 < M < \tilde{M} \\[2ex] \dfrac{2M}{M^{*2}} \propto \dfrac{M}{\xi^2 l^2} \,, & \tilde{M} < M < M^* \\[2ex] 0 \,, & M < M_1 \ or \ M > M^* \end{cases} \qquad (3.7)$$

where $M^*/\tilde{M} = l^\gamma$, with $\gamma \approx 1$ for large α and $M_1 = 2\sigma a/(2l^2+1)$ is the moment of single block event. The fact that (3.7) has explicit l dependence means that the "ultraviolet cutoff" $a = \xi/l$ plays an important role. For real faults, such a small length scale may result from stable creep [9]. If the sizes of the smallest earthquakes are of the order of tens of meters [11], then the natural value for l is of order 10^2.

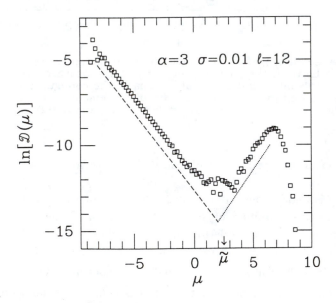

Fig. 6. The frequency-magnitude distribution. The "magnitude" $\mu = \ln M$. Note that the localized events and delocalized events have different statistics. The crossover is roughly at $\tilde{\mu} = \ln \tilde{M}$ which is indicated by the arrow. The dashed line has a slope -0.92; while the slope of the dotted line is 1. The parameters used in the simulation are shown in the figure.

The average displacement of blocks must be the same as that of the plate to which they are attached, i.e.,

$$\int_{M_1}^{M^*} M \mathscr{D}(\mu) \, d\mu = 1.$$

(3.8)

The contribution of localized events $(M < \tilde{M})$ to the integral (3.8) can be estimated; it is of the order of $\frac{1}{l^2}\ln\frac{l^3}{\sigma}$, which is very small for $l \approx 100$ (assuming that σ does not play a crucial role). Almost all the contributions to (3.8) come from the delocalized events with $M \sim M^*$.

4. DISCUSSION

Earthquake models with many degrees of freedom operating at or near a threshold instability exhibit very rich behaviors. The fluctuations (earthquakelike events) cover a wide range. Both models show some scaling behavior. If the above version of the Burridge-Knopoff (BK) model is appropriate for a single fault, then the motion of a single fault is dominated by either very large or system wide events depending on whether the size of M^* is smaller or larger than the system size. There is a scaling region in this model for small events with $B \approx 1$, and this B-value seems to be one of the most robust features of the model. It would be very interesting and important to see if real seismological data for small earthquakes (which is difficult to collect and analyze) scale with $B \approx 1$. It may well be that each fault behaves like the BK model, and the Gutenberg-Richter law (1.1) with $B \approx 2/3$ results from the interaction (or distribution) of faults. In another words, (1.1) is a property of "geometry" of faults. Of course, this geometry is generated dynamically by the tectonic motion. The results of CHEN et al. [5] are certainly suggestive. However, certain key issues remain to be resolved. For example, the B-value of their model for two dimensions is 0.4. This would imply a crossover to a smaller B-value (than 2/3) for large earthquakes which are two-dimensional-like. Such a crossover in real data remains to be seen. More importantly, what happens if some of the more realistic, but seemingly essential features (e.g., the redistribution of stress propagates, instead of infinitely fast, at the speed of sound so that the causality is preserved) are added to the model? How to incorporate the scaling law

$M_0 \sim A^{3/2}$ (A is the rupture area) [12] in the model? The total number of sites s which have ruptured in an event is used to estimate M_0, but s seems to be more close to A than to M_0.

I thank K. Chen for providing Figs. 2 and 3.

REFERENCES

1. B. Gutenberg and C. F. Richter, Ann. Geofis. **9**, 1 (1956).
2. P. Bak, C. Tang, and K. Wiesenfeld, Phys. Rev. Lett. **59**, 381 (1987); Phys. Rev. A **38**, 364 (1988). C. Tang and P. Bak, Phys. Rev. Lett. **60**, 2347 (1988); J. Stat. Phys. **51**, 797 (1988).
3. See, for example, C. H. Scholz, *The Mechanics of Earthquakes and Faulting*, (Cambridge, 1990) for references.
4. P. Bak and C. Tang, J. Geophys. Res. **94**, 15635 (1989); also see P. Bak, C. Tang, and K. Wiesenfeld, in *Cooperative Dynamics in Complex Systems*, edited by H. Takayama (Springer, Tokyo, 1988), p. 274.
5. K. Chen, P. Bak, and S. P. Obukhov, Phys. Rev. A **43**, 625 (1991).
6. P. Bak and K. Chen, in *Fractals and Their Application to Geology*, edited by C. Barton and P. LaPointe (Geological Society of America, Denver, 1991).
7. R. Burridge and L. Knopoff, Bull. Seismol. Soc. Am. **57**, 341 (1967).
8. For example, see J. Rundle and D. Jackson, Bull. Seismol. Soc. Am. **67**, 1363 (1977); A. Nur, Pageoph. **116**, 964 (1978); T. Cao and K. Aki, *ibid.* **122**, 10 (1984); H. Takayasu and M. Matsuzaki, Phys. Lett. A **131**, 244 (1988).
9. J. M. Carlson and J. S. Langer, Phys. Rev. Lett. **62**, 2632 (1989); Phys. Rev. A **40**, 6470 (1989).
10. J. M. Carlson, J. S. Langer, B. Shaw, and C. Tang, Phys. Rev. A **44**, 884 (1991).
11. P. E. Malin, S. N. Blakeslee, M. G. Alvarez, and A. J. Martin, Science **244**, 557 (1989).
12. H. Kanamori and D. L. Anderson, Bull. Seismol. Soc. Am. **65**, 1073 (1975).

Application of a Mean Field Approximation to Two Systems that Exhibit Self-Organized Criticality

J. Theiler

In this exposition, a mean field analysis will be applied to predict and explain some of the features observed in two systems that are known to exhibit self-organized criticality: the sandpile model of BAK, TANG, and WIESENFELD [1], and a variation using continuous values that was introduced by ZHANG [2]. It will be argued that mean field of TANG and BAK [3] is problematic in that it fails to converge. The modification suggested here introduces a parameter to account for sandgrains falling off the edge of the sandpile; this balances the sandgrains which are dropped on the sandpile from above. The modified analysis is then applied to the equilibrium state of the sandpile, and to the time evolution toward equilibrium. The analysis is then extended to other systems which exhibit self-organized critical behavior.

1. Introduction

For dissipative dynamical systems with extended degrees of freedom, a variety of apparently collective phenomenon have been observed, including spatio-temporal chaos [4], robust intermittency [5], long quasistationary transients [6], attractor crowding [7] and clustering [8, 9], "amplitude death" [10], and self-organized criticality [1]. Self-organized criticality, in particular, has been invoked to explain behavior of a wide variety of physical systems, from earthquakes to avalanches.

The mean field is one of the few analytical tools that are available for the understanding of systems with many interacting components. The approach of the mean field is to replace individual components with statistical averages. It is a relatively crude approximation which ignores the spatial aspect of the system, but it is general enough to be applied to many different systems, and it usually leads to a computationally tractible approximation.

After introducing the sandpile model, I will discuss the the mean field analysis that was developed by TANG and BAK [3], and show how a slight modification makes the analysis more sensible. The modified analysis will then be compared to simulations and the quantitative agreement (and disagreement) will be discussed. Finally, the analysis will be extended to the continuous system of ZHANG [2], and it will be shown that the

J. Theiler: MIT Lincoln Laboratory; *and* Center for Nonlinear Studies and Theoretical Division, MS-B213, Los Alamos National Laboratory, Los Alamos, NM 87545.
E-mail: jt@infidel.lanl.gov

self-organized discretization which this system displays can be understood in terms of a mean field.

2. The sandpile model

The sandpile model was introduced by BAK, TANG, and WIESENFELD [1], hereafter BTW, to illustrate a property they called self-organized criticality. At each site (x, y) on a square lattice, there is an integer number $z(x, y)$ of what are often, but not quite correctly, called sandgrains. (It would be only slightly more correct to call $z(x, y)$ a "local slope," but "sandgrains" provide a more concrete picture.] There is a critical value $K = 4$ above which a lattice site is unstable. If any site on the lattice is unstable, then the lattice as a whole is considered unstable, and is permitted to relax in a series of discrete steps. Four sandgrains at each unstable site are redistributed, one to each of the four neighboring sites, in a kind of diffusion that is highly nonlinear. Here,

$$
\begin{aligned}
z(x, y) &\rightarrow z(x, y) - 4, \\
z(x, y \pm 1) &\rightarrow z(x, y \pm 1) + 1, \\
z(x \pm 1, y) &\rightarrow z(x \pm 1, y) + 1.
\end{aligned}
\tag{1}
$$

The relaxation process continues until every site on the lattice is stable.

The dynamics then proceeds by seeding at a single random (or a fixed [11]) site: $z(x_o, y_o) \rightarrow z(x_o, y_o) + 1$. If this leads to an unstable lattice, the lattice is allowed to relax by the above rule until it becomes stable again. Thus there are "long" time steps between each seeding, and "short" time steps, during which the lattice relaxes.

In simulations, one usually waits until the lattice is finished relaxing until the next seeding; the point of this is to ensure that the two time scales (slow seeding and fast relaxing) are fully separated. The physical process which does the seeding presumably is doing so at a slow rate that is independent of the state of the lattice.

2.1. Notation

The lattice in the BTW system needn't be two dimensional, nor need it even be a lattice. Let $\mathbf{x}$ represent a site on the lattice, and $z(\mathbf{x})$ be the value at that site. Each site is connected to its neighbors $\mathbf{x}' \in \mathcal{N}_{\mathbf{x}}$. (Note that there is a bi-directionality in the case of a lattice: $\mathbf{x}' \in \mathcal{N}_{\mathbf{x}}$ implies $\mathbf{x} \in \mathcal{N}_{\mathbf{x}'}$.) It is possible to represent the external seeding by a field term $h_t(\mathbf{x})$, which is zero almost all of the time, but equal to one when the $\mathbf{x}$ is seeded. (The t here corresponds to short time steps.) Further, write $\xi_t(\mathbf{x})$ as the "backflow" from neighboring sites. Let $b = |\mathcal{N}_{\mathbf{x}}|$ be the number of bonds per site (also called the "coordination number"), and let K be the threshold value (note $b = K = 4$ for the original two-dimensional BTW system).

$$
\begin{aligned}
z_{t+1}(\mathbf{x}) &= z_t(\mathbf{x}) + h_t(\mathbf{x}) - bF(z_t(\mathbf{x})) + \xi_t(\mathbf{x}) \\
\xi_t(\mathbf{x}) &= \sum_{\mathbf{x}' \in \mathcal{N}_{\mathbf{x}}} F(z_t(\mathbf{x}'))
\end{aligned}
\tag{2}
\tag{3}
$$

where $F(z)$ is the threshold function: its value is 0 for $z < K$ (inactive), and 1 for $z \geq K$ (active).

3. The mean field of Tang and Bak

In the mean field introduced by TANG and BAK [3], hereafter TB, the approximation is made that the sites are statistically independent and identically distributed. The site value z is treated as a random variable whose evolution is given by the equation

$$z_{t+1} = z_t - bF(z_t) + h_t + \xi_t \tag{4}$$

where h_t is a random variable that takes a value of 1 with probability h and a value of zero the rest of the time. The backflow, ξ_t, is a random variable given by

$$\xi_t := \sum_{i=1}^{b} F(z_t^{(i)}) \tag{5}$$

where $z_t^{(i)}$ are random variables that correspond to the neighbors of z_t. By the assumption of the mean field, these are taken to be independent but with the same distribution as z_t. Now ξ_t takes on a value of x when exactly x of the site's neighbors are active. Thus,

$$\mathcal{P}\{\xi = x\} = \binom{b}{x} A^x I^{b-x} \tag{6}$$

where, following TB, I have written $A := \mathcal{P}\{F(z) = 1\} = \mathcal{P}\{z \geq K\}$ as the probability that a site is active, and $I := 1 - A = \mathcal{P}\{z < K\}$ as the probability that a site is inactive.

Now, I can write a transition matrix for the Markov process. Let $\mathcal{T}_t(z \leftarrow z') = \mathcal{T}_t[z][z'] := \mathcal{P}\{z_{t+1} = z \mid z_t = z'\}$ be the probability of a transition at the fiducial site from a value of z' to a value of z. In general, I have

$$\mathcal{T}_t[z][z'] = \begin{cases} \mathcal{P}\{h_t + \xi_t = z - z'\} & \text{for } z' < K, \\ \mathcal{P}\{h_t + \xi_t = z - z' + b\} & \text{for } z' \geq K, \end{cases} \tag{7}$$

which we can write $\mathcal{T}_t[z][z'] = \mathbf{T}_t[z - z' + bF(z')]$ where the transition vector $\mathbf{T}$ is defined by

$$\begin{aligned} \mathbf{T}[x] &:= \mathcal{P}\{h + \xi = x\} \tag{8} \\ &= (1-h)\mathcal{P}\{\xi = x\} + h\mathcal{P}\{\xi = x - 1\} \tag{9} \\ &= (1-h)\binom{b}{x} A^x I^{b-x} + h\binom{b}{x-1} A^{x-1} I^{b-x+1}. \tag{10} \end{aligned}$$

The transition matrix $\mathcal{T}_t$ allows us to evolve the vector of current probabilities $\mathbf{P}_t[z] := \mathcal{P}\{z_t = z\}$. We have, in matrix notation, $\mathbf{P}_{t+1} = \mathcal{T}_t \mathbf{P}_t$, or more explicitly

$$\begin{aligned} \mathbf{P}_{t+1}[z] &= \sum_{z'} \mathcal{T}_t[z][z'] \mathbf{P}_t[z'] \tag{11} \\ &= \sum_{z'} \mathbf{T}_t[z - z' + bF(z')] \mathbf{P}_t[z'] \tag{12} \end{aligned}$$

Thus, we can derive the time evolution of the probability distribution of the random variable z. Because the the matrix $\mathcal{T}_t$ depends on the vector $\mathbf{T}_t$, which depends on A_t and I_t, both of which in turn depend on $\mathbf{P}_t$, one can think of the evolution of probabilities in Eq. 12 as a nonlinear map:

$$\mathbf{P}_{t+1} = f(\mathbf{P}_t). \tag{13}$$

3.1. The problem of nonconvergence

The problem with this version of the mean field is that it does not converge. For any $h > 0$, there is no solution $\mathbf{P}_\infty[z]$ which satisfies $\mathbf{P} = f(\mathbf{P})$. In particular, it is straightforward to show that for this model, the average value $\langle z_t \rangle := \sum_z z \mathbf{P}_t[z]$ does not converge. Instead, we have, from Eqs. (4) and (5), that

$$\langle z_{t+1} \rangle = \langle z_t \rangle + h. \tag{14}$$

This is in retrospect not really surprising, since the dynamics preserves the total number of "sandgrains" (each sand grain lost at site $\mathbf{x}$ is picked up again at a neighboring site $\mathbf{x}'$), and a steady field h is being added at each step. In an actual BTW simulation, this is not a problem, since avalanches occasionally carry those added grains of sand off the edge.

4. Modified mean field: introducing edges

One solution to the problem of nonconvergence is to mimic what happens in an actual BTW lattice, and to let sandgrains fall off the edge. It isn't instantly obvious how to do this, since a mean field approximation by definition eliminates the lattice structure from the problem, and fails to distinguish edges from interior sites. The suggestion here is to state that a fraction $e > 0$ of the sites are on the edge; we can later take e very small if that is desired, but we need some mechanism to "lose" those grains of sand that are being added by the field. Having a finite fraction of edge sites alters the equation for the backflow ξ. It is still the case that ξ is equal to the number of active neighbors, but the number of available neighbors depends now on whether the site is on the boundary or not. With probability e it is on the boundary, and has $b - 1$ available neighbors; with probability $1 - e$ it is in the interior, and has b available neighbors. Thus,

$$\mathcal{P}\{\xi = x\} = (1 - e)\binom{b}{x} A^x I^{b-x} + e\binom{b-1}{x} A^x I^{b-1-x}. \tag{15}$$

and the modified transition vector is given by

$$
\begin{aligned}
\mathbf{T}_t[x] &= (1 - h)\left[(1 - e)\binom{b}{x} A_t^x I_t^{b-x} + e\binom{b-1}{x} A_t^x I_t^{b-1-x}\right] \\
&+ h\left[(1 - e)\binom{b}{x-1} A_t^{x-1} I_t^{b-x} + e\binom{b-1}{x-1} A_t^{x-1} I_t^{b-1-x}\right].
\end{aligned}
\tag{16}
$$

The first thing to note is that $e = 0$ leads to the original TB mean field. However, in this case, we have that the average evolves according to

$$\langle z_{t+1} \rangle = \langle z_t \rangle + h - eA_t. \tag{17}$$

Thus, a steady state can be achieved, with the balance $h = eA$ corresponding to external input on the left hand side, and loss over the edge on the right hand side. Note that this balance equation does not make sense in the original model of TB, where $e = 0$.

4.1. Finding the self-consistent solution

Having defined the modified mean field equations (12) and (16), it is straightforward to solve for the probabilities $\mathbf{P}$. Numerically, one need only evolve the probabilities forward in time until a steady state is reached. The solution exists for nonzero h and e as long as $h < e$.

In particular, it is possible to show [12] that in the simultaneous limit $h \to 0$ and $e \to 0$, with $h/e \to A$

$$\text{for } 0 \le z < b, \quad \mathbf{P}[z] = (1/b) \sum_{k=0}^{z} \binom{b}{k} A^k (1-A)^{b-k} \tag{18}$$

$$\text{for } b \le z < 2b, \quad \mathbf{P}[z] = (1/b) \sum_{k=z-b}^{b} \binom{b}{k} A^k (1-A)^{b-k} \tag{19}$$

$$\text{for } z > 2b, \quad \mathbf{P}[z] = 0. \tag{20}$$

This leads to the result

$$\langle z \rangle = (b-1)/2 + bA, \tag{21}$$

which although linear in A, is in fact is exact for $0 \le A < 1$. The sandpile model does not re-seed until the lattice is finished relaxing; this is equivalent to saying $h \to 0$ for fixed, but arbitrarily small, e. Thus, the $A \to 0$ limit is appropriate as an approximation of the sandpile model. This in fact is the same equilibrium probabilities that TB claim to get in the $h \to 0$ limit for their analysis.

4.2. Some comments on computing exponents

In TB, four equations are presented, Eqs. (4a-d) in their paper, with four unknowns (P_0, P_1, P_2, P_3) and one parameter (h). A second parameter $(\theta$, which is the same as $\langle z \rangle)$ is introduced as a function of the four variables. All of these are combined into a single equation, Eq. (5), which appears to exhibit two degrees of freedom; that is, both h and θ are treated as though they were free to vary independently of each other. From this, a variety of exponents are computed. But h and θ are not independent. As OBUKHOV [13] has pointed out, simply adding a field h brings the system away from from the critical point, so that "the parameters which describe both the proximity to the critical point and the magnetic field are coupled together." However, I would argue that not only are there not two independent degrees of freedom; there is not even one. For while there is complete degeneracy in the solution when $h = 0$ (constrained only by $\mathbf{P}[0] + \cdots + \mathbf{P}[b-1] = 1$ and $\mathbf{P}[b] = \mathbf{P}[b+1] = \cdots = 0$), taking h to any nonzero value leads to inconsistent equations with no solution at all. The edge parameter e provides one mathematically valid way to break the degeneracy.

For the mean field analysis presented here, the edge parameter is considered fixed, so there is really only one degree of freedom, in the control parameter h. So I cannot compute exponents as they are defined in TB. Because of this, I have concentrated not on predicting exponents (even though these are arguably the most interesting features), but rather the relative frequencies of the values on the lattice, $\mathbf{P}[z]$, and in particular, the average $\langle z \rangle$. I should comment that there are other approaches which also go by the name "mean field" but which are based on treating an avalanche as a branching process; these are discussed by ALSTRØM [14] and OBUKHOV [13]. Unlike the mean field analysis presented here, these approaches *are* able to predict nontrivial exponents.

4.3. Comparison with Simulation

The mean field analysis predicts $\mathbf{P}[x] = 1/b$ for $x < b$ and $\mathbf{P}[x] = 0$ for $x \ge b$, giving $\langle z \rangle = (b-1)/2$. It is of obvious interest to compare these predictions with values obtained from numerical simulations of the BTW sandpile. The assumption of the mean field is

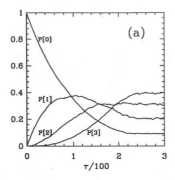
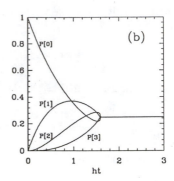

Figure 1: **(a)** An average of 50 simulations of a 10x10 lattice, showing the probabilities $P_t[z]$ for the stable values $z = 0, 1, 2, 3$ as a function of time τ scaled by the area of the lattice $S = 100$. The initial condition was for all sites to be zero. **(b)** Evolution of the mean field equations, with time t scaled by the field parameter h.

that is that sites are statistically independent. Disagreements between mean field theory and simulations therefore point to the importance of site-to-site correlations.

For the one dimensional sandpile (with $b = K = 2$) the sandpile self-organizes into a state in which almost all of the sites are minimally stable. In the limit of large lattice size, $\mathbf{P}[1] \to 1$, and $\langle z \rangle \to 1$. By contrast, the mean field predicts $\mathbf{P}[0] = \mathbf{P}[1] = 1/2$, and $\langle z \rangle = 1/2$.

A two dimensional square lattice has $b = K = 4$, and the mean field predicts $\mathbf{P}[0] = \mathbf{P}[1] = \mathbf{P}[2] = \mathbf{P}[3] = 1/4$, and $\langle z \rangle = 1.5$. However, precise numerical simulations by Manna [15] find $\mathbf{P} = (0.073, 0.174, 0.307, 0.446) \pm 0.003$ and $\langle z \rangle = 2.124$.

Because the mean field is better at predicting the equilibrium state of a two dimensional system than a one dimensional system, one is led to presume that it will be even better for higher dimensions. It is often the case for critical phenomena that there is an upper critical dimension, beyond which all behavior is independent fo dimension and depends only on coordination number. Indeed, a number of authors have suggested that there is an upper critical dimension for the sandpile [2, 3, 13, 14, 16, 17]. However, these authors are concerned with the critical exponents; I do not know of work which suggests that the individual probabilities $\mathbf{P}$ will be exact for lattices above a critical dimension.

4.4. Time evolution

The mean field analysis not only predicts a self-consistent equilibrium state, but also the time evolution of the system as it evolves toward equilibrium. In comparing the mean field to simulations on the BTW sandpile, however we must first make sure that equivalent notions of "time" are used.

The t in the mean field approximation refers to the short time steps during which the lattice is relaxing. Since h is the probability of seeding a single site at a given short time step, $1/h$ is the average number of short time steps between seedings of a particular site. If there are a total of S sites, then $1/(hS)$ is the average number of short time steps between each seeding of the whole lattice; that is, the length of the long time step. Then, $\tau = hSt$ is time measured in units of long time steps.

In Fig. 1(a), the time evolution of a 10×10 sandpile lattice is simulated, starting

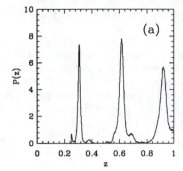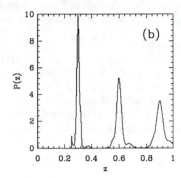

Figure 2: **(a)** Simulation of the Zhang model on a two-dimensional 50x50 square lattice. Not shown is the delta function peak at $z = 0$. **(b)** Mean field approximation for the Zhang model with $e = 0.05$.

with the initial condition of $z(x, y) = 0$ for every site. Plotted are the frequencies of occurrence of site values 0, 1, 2, and 3. The time plotted is number long time steps divided by the lattice size: $\tau/S = \tau/100$.

In Fig. 1(b), the mean field is evolved using the evolution equation Eq. 13 with very small h. Here, the time axis is ht where t is the short time step in Eq. 13. The mean field model is in good agreement with the simulation for short times, and does reasonably well predict the time required for equilibrium to be reached.

5. Continuous Lattice system of Zhang

A variation of the sandpile model of BTW was proposed by ZHANG [2]. This model uses continuous instead of discrete values z and whenever the value at a site exceeds a critical value K, the site relaxes according to the rule

$$
\begin{aligned}
z(x, y) &\rightarrow 0 \\
z(x \pm 1, y) &\rightarrow z(x \pm 1, y) + z(x, y)/4 \\
z(x, y \pm 1) &\rightarrow z(x, y \pm 1) + z(x, y)/4
\end{aligned}
\tag{22}
$$

Seeding is done by addition of a random input uniformly chosen in the range $[0, 2K/b]$ (so that the average input seed is equal to K/b which corresponds to the BTW system) at a random site. Zhang notes that, like the BTW sandpile, this system also exhibits self-organized criticality.

Another kind of self-organization is also observed. After a period of evolution, one finds that a histogram of z values exhibits distinct peaks. For a two dimensional square lattice, there are four such peaks, which is just how many distinct states are available to a site in the original BTW sandpile. (See Fig. 2(a)) One of the peaks is a delta function at $z = 0$, and the others are sharp but of finite width.

While leaving the details to a forthcoming paper [12], I comment that the modified mean field analysis which was used in §4. for the sandpile model is readily adapted to Zhang's continuous model. As Fig. 2(b) shows, the mean field is able to predict the discretization observed in the simulations. As in the case of the BTW sandpile, the details of the distribution of $P(z)$ are only approximately captured. Further experiments [12] indicate that the widths of the peaks can be attributed to the finite size of the lattice; they are made sharper in the mean field approximation by decreasing the edge parameter e.

This work was initiated during a visit to the Complex Systems Group (T-13) at Los Alamos National Laboratory; work at MIT Lincoln Laboratory was sponsored by the United States Air Force and the Defense Advanced Research Projects Agency (DARPA). Further work continued at Los Alamos under the auspices of the Department of Energy. I am grateful to Bill Bruno, Bette Korber, Wentian Li, and Kurt Wiesenfeld for useful discussions, and I particularly want to thank Lui Lam for his enthusiasm and generosity before and during this excellent conference.

References

1. P. Bak, C. Tang, and K. Wiesenfeld. Self-organized criticality: An explanation of $1/f$ noise. *Phys. Rev. Lett.* **59**, 381 (1987).
2. Y.-C. Zhang. Scaling theory of self-organized criticality. *Phys. Rev. Lett.* **63**, 470 (1989).
3. C. Tang and P. Bak. Mean field theory of self-organized critical phenomena. *J. Stat. Phys.* **51**, 797 (1988).
4. J. P. Crutchfield and K. Kaneko. Phenomenology of spatio-temporal chaos. In Hao Bai-Lin, editor, *Directions in Chaos, Vol. I*. World Scientific, pages 272–353, 1988.
5. J. D. Keeler and J. D. Farmer. Robust space-time intermittency and $1/f$ noise. *Physica D* **23**, 413 (1986).
6. J. P. Crutchfield and K. Kaneko. Are attractors relevant to turbulence? *Phys. Rev. Lett.* **60**, 2715–2718 (1988).
7. K. Wiesenfeld and P. Hadley. Attractor crowding in oscillator arrays. *Phys. Rev. Lett.* **62**, 1335 (1989).
8. K. Kaneko. Clustering, coding, switching, hierarchical ordering, and control in a network of chaotic elements. *Physica D* **41**, 137–172 (1990).
9. L. Fabiny and K. Wiesenfeld. Clustering behavior of oscillator arrays. *Phys. Rev. A* **43**, 2640–2648 (1991).
10. R. E. Mirollo and S. H. Strogatz. Amplitude death in an array of limit-cycle oscillators. *J. Stat. Phys.* **60**, 245–262 (1990).
11. K. Wiesenfeld, J. Theiler, and B. McNamara. Self-organized criticality in a deterministic automaton. *Phys. Rev. Lett.* **65**, 949–952 (1990).
12. J. Theiler. Mean-field analysis of systems that exhibit self-organized criticality. Working paper, 1991.
13. S. P. Obukhov. The upper critical dimension and ϵ-expansion for self-organized critical phenomena. In *Random Fluctuations and Pattern Growth: Experiments and Models*, H. E. Stanley and N. Ostrowski, eds., volume 157 of *NATO Advanced Science Institutes, Series E*, Kluwer, Dordrecht, The Netherlands, pages 336–339, 1988.
14. P. Alstrøm. Mean field exponents for self-organized critical phenomena. *Phys. Rev. A* **38**, 4905–4906 (1988).
15. S. S. Manna. Large-scale simulation of avalanche cluster distribution in sand pile model. *J. Stat. Phys.* **59**, 509–521 (1990).
16. D. Dhar and R. Ramaswamy. Exactly solved model of self-organized criticality. *Phys. Rev. Lett.* **63**, 1659–1662 (1989).
17. T. Hwa and M. Kardar. Dissipative transport in open systems: An investigation of self-organized criticality. *Phys. Rev. Lett.* **62**, 1813 (1989).

Part V
Fluids and Crystal Growths

Modeling the Hydrodynamics of Materials Processing

W.E. Langlois

During the past decade, fluid dynamics has emerged as a useful quantitative tool for understanding the physical phenomena that operate in materials processing. One of the most important examples is crystal growth from the melt. The lecture will take this as an paradigm, to show how digital simulation has been used to obtain insight into these complicated situations.

1. INTRODUCTION

One of the most prominent features of a moving fluid is advective transport. Various physical quantities — heat, momentum, solutes — are carried from point to point by the velocity field. By contrast, except in the rare cases when internal radiative heat transfer is important, transport within a solid is carried out almost exclusively by diffusion. Elastic and plastic deformations of a solid do move things about, but not very far, and advection is usually unimportant.

Of course, things diffuse within fluids too. The net transport is determined both by advection and diffusion, and, in the case of heat, perhaps also by radiation. If transport is important, all forms must be considered unless one or more can be ruled out as negligibly small.

A brief digression about terminology: In some branches of fluid dynamics, the mech-

W.E. Langlois: IBM Research Division, Almaden Research Center, 650 Harry Road San Jose, CA 95120-6099, USA

anism I've been calling "advection" is instead called "convection", and in others the two terms are used more-or-less synonymously. However, when thermodynamic processes are involved, and in materials processing they usually are, it is useful to give the terms separate and distinct meanings. "Advection" simply means the transport that arises because things are carried about by the moving fluid, regardless of what is making it move. "Convection" is reserved for motions that arise from gradients of temperature or of solute concentration.

To return to the main topic, many aspects of materials processing involve solidification from a fluid phase. Casting, extrusion, crystallization, calendering, welding, soldering, and chemical vapor deposition are some examples which do not exhaust the list.

As recently as the 1960's, most theoretical investigations of these processes were based on the assumption that heat and mass transport in the fluid phase are diffusion dominated. The reason for this is that people doing research continue to remain people. Most of us, when we lose a valuable object, tend to search first where the light is good. If that doesn't work, we get down on the floor and grope around in the dark.

The analogy here is that diffusion is usually a linear process involving material coefficients that are relatively easy to measure. Thus, the entire toolbox of linear mathematics is available to the theoretician, and he has seldom has a problem getting good numbers to put into his model.

Unfortunately, in many materials processing applications, perhaps the majority, advection is as important as diffusion -- often, in fact, more important. Whenever solving for the fluid motion is part of the problem, advection is nonlinear. Beginning in the early 1960's computational methods for dealing with nonlinear problems emerged rapidly, but the problem of model coefficients remained. For materials like molten metals and binary mixtures, parameters such as viscosity and volumetric expansion coefficient are not easy to measure. What measurements did exist were not widely disseminated.

During the past quarter century, however, the importance of advection in materials processing has come to be appreciated. The period has seen major advances in our understanding of the hydrodynamics involved. Many aspects were reported in the individual papers of [1].

Instead of attempting to survey all the processes in which hydrodynamics plays a role, I shall use one as a paradigm, viz. growth of crystals from the melt by a specific method known as Czochralski growth. This system has been extensively studied for several reasons. First of all, it is the process by which some important optical and electronic materials, including silicon, are grown. Also, the geometry of the process is relatively simple, which made it feasible for modelers to incorporate a high level of physical complexity. Finally, transport coefficients for the most important material, silicon, have now been measured to a good degree of accuracy.

2. A DESCRIPTION OF CZOCHRALSKI GROWTH

In the Czochralski process, pictured in Fig. 1, a crucible is initially loaded with the material to be crystallized. When this is melted, a seed crystal held at the end of a pull rod is dipped in. After appropriate start-up procedures, the growing crystal is slowly extracted; pull rates in the range of millimeters or centimeters per hour are typical.

To keep the nutrient molten, the crucible is maintained at a temperature above the freezing point. The resulting radial temperature gradient is itself sufficient to cause convection in the melt. Other driving forces complicate matters. The growing crystal is some-

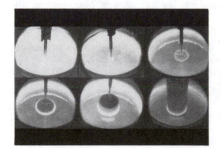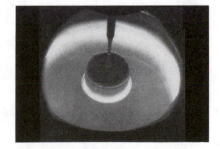

Fig. 1. Photographs taken inside a Czochralski furnace. The sequence in the left half shows various stages of the process, from (top left) the dipping of the seed though the last part of the run. The right half is a close-up of an intermediate stage.

times rotated as it is pulled. This contributes a rotary and centrifugal motion intended to produce desirable conditions near the growth interface. Also in some applications, including silicon growth, the crucible also is rotated, to smooth out thermal asymmetries that arise from irregularities in the heating. This too generates a centrifugal flow. Moreover, it imparts rotational velocity to most of the melt mass, leading to the non-intuitive gyroscopic effects found in rotating fluid systems. In particular, eddies that arise within the melt show a pronounced tendency to form themselves the vertical structures known as Taylor columns.

There is still one more driving mechanism that is much less obvious. In fact, many hydrodynamicists work their entire careers never needing to deal with it. That mechanism is thermocapillary convection, which occurs in a fluid mass with a free surface when temperature variation leads to a gradient of the surface-tension coefficient within the surface. Recent interest in hydrodynamic experiments in space vehicles has kindled interest in thermocapillarity convection. In conditions of very low gravity, buoyant convection mostly goes away. As a result, thermocapillary convection often dominates the motion, even if it would be negligible under normal gravity conditions. In fluid motions where there is a temperature gradient in the free surface, and a strong dependence of surface tension on temperature, thermocapillary convection can be significant even in normal gravity. Both these conditions are met in Czochralski growth.

When all of these driving mechanisms act in concert, the resulting flow can be quite complicated. The correspondingly complicated advective transport of heat, solutes, and impurities toward the growth interface can have a profound influence on the properties of the finished crystal, and is therefore of importance to the crystal grower. In certain parameter ranges, which do occur in practice, the motion can lose its axial symmetry or even become turbulent. However, the crystal grower usually strives to keep everything as smooth and symmetric as possible. Hence the regime of laminar axisymmetric flow is important in practice.

The importance of hydrodynamics in crystal growth began to be appreciated in the mid-1960's. Flow visualization studies carried out at or near room temperature demonstrated that complicated motions can occur in crystal-growth geometries, and that qualitatively different regimes of flow can be triggered by varying external parameters such as rotation rates or temperature differences.

Crystal growth flows have also been investigated using the methods of computational fluid dynamics, and a number of computer models have been developed. Unlike the flow visualization studies, these are not confined to low temperature investigations. Consequently, they brought to light even more complicated phenomena. For example, oscillatory flow fields were sometimes found to occur, even when all external effects were steady, establishing that hydrodynamics can be one of the causes of striations in crystals. Also, thermocapillary convection was confirmed as important, even in large-scale terrestrial crystal growth systems. For some time, however, accurate values of the transport coefficients were still hard to come by. The computations, even when carried out with well-tested programs on high resolution grids, therefore gave only semi-quantitative results. This phase of the work was reviewed in [2].

During the last few years, the absence of accurate parameter data has been somewhat corrected. Silicon, in particular, has been well characterized. This has made it possible to carry out quantitative studies of fluid flow in this important melt using the computer models. Comparison of the flow patterns predicted by the models with those occurring in actual melts is not feasible, because of the difficulties involved in observing the latter. However, incorporating solute transport into a model allows it to make predictions which can be checked against experiment. For example, the patterns of solute deposition into growing crystals can be computed. Since the patterns in actual crystals can be quite complex, and the physical mechanisms leading to them even more so, the satisfactory agreement between simulation and experiment [3, 4] gives credibility to present-day computational models. Working in concert with careful experimental studies and asymptotic approaches, they can be valuable tools for quantitative understanding of crystal growth hydrodynamics.

3. COMPUTATIONAL FORMULATION OF CZOCHRALSKI HYDRODYNAMICS

Simulating a crystal growth flow requires us to deal with a moderately complicated system of partial differential equations and boundary conditions. To see how they arise, let us begin with a survey of the underlying fluid dynamics. For those having a background in continuum mechanics, much of this will be review. However, there are points of interpretation, important for simulations of crystal growth flows, that are not always stressed in treatments of fluid dynamics.

It would seem that the work could be divided neatly into two stages. In the first, one derives the equations and boundary conditions from the underlying physical principles. In the second, one accepts them as "given", and proceeds to address them as purely mathematical entities.

However this is a poor way to proceed, because the physical principles remain important even during the computational stages of the work. There are many ways to truncate the equations for numerical computation, but not all are physically acceptable. It is easy, for example, to truncate in a way that leads to false production of angular momentum, with disastrous results.

The remainder of this section reviews the physical principles and the way they lead to a computational formulation. Their relevance to numerical simulation is discussed in context.

3.1 The Balance Laws

The central equations of hydrodynamics are really statements that certain physical quantities (mass, momentum, etc.) are neither created nor destroyed. Obtaining the mathematical forms of the statements from the verbal forms is an exercise in elementary vector calculus.

One can, for example, focus attention on a region of space immersed in the fluid. At any instant of time, the total amount of some variable within this region is obtained by integrating the density of the variable over the region. For a conserved quantity, the time rate of change of this amount is the net source acting within the region (another volume integral), augmented by the net flux across the boundary of the region (a surface integral) and by sources acting at the boundaries (another surface integral). Gauss's divergence theorem is then used to convert the surface integrals to volume integrals. Transposing everything to the same side of the equation leads to a statement that a certain integral over the volume vanishes. Since the statement holds for any volume immersed in the fluid, the integrand itself vanishes identically.

Applying this procedure to the fluid mass, which is relatively easy because there are no internal sources, leads to the familiar *equation of continuity*

$$\partial \rho / \partial t + \nabla \bullet (v\rho) = 0 , \tag{1}$$

where ρ denote the fluid density and v denotes the velocity.

The case of momentum is slightly more complicated because impulses associated with the body force $\mathbf{F}$ and the stress $\mathbf{T}$ act as sources for momentum. The density of momentum in the x_i direction is ρv_i. Applying the procedure leads to an equation governing its time evolution:

$$\partial(\rho v_i)/\partial t + \nabla \bullet (v\rho v_i) = \rho F_i + \partial T_{ji}/\partial x_j , \tag{2}$$

where the repeated subscript j in the last term denotes summation over j = 1,2,3.

At this point the equation of continuity could be used to simplify (2), reducing it to

$$\partial v_i / \partial t + (\mathbf{v} \bullet \nabla)v_i = F_i + \rho^{-1} \partial T_{ji}/\partial x_j . \tag{3}$$

For analytical work, this is often useful. However we have come to the first point where numerical work is different. In essence, constructing a finite difference scheme reverses the procedure that led to (1) and (2). Partial derivatives are replaced by finite analogs, leading to equations for the grid-cell averages of the dependent variables. Using (2) rather than (3) makes it easier to do this in a way which keeps accurate track of the momentum flux. This is important: since the three components of momentum are physically conserved quantities, momentum advected out of a grid cell should be exactly matched by an increase in the momentum inventory of neighboring cells. It is especially crucial to use such *conservative differencing schemes* in closed systems such as crystal growth flows, because spurious momentum will not be swept "downstream" — there is no "downstream". When the production and advection of a variable f are written in the form

$$\partial f/\partial t + \nabla \bullet (vf) ,$$ (4)

the resulting equation is said to be in *flux form* or *conservation form*.

The flux form derived from conservation of total energy is

$$(\partial/\partial t)[\rho(\tfrac{1}{2}v \bullet v + e)] + \nabla \bullet [v\rho(\tfrac{1}{2}v \bullet v + e)]$$
$$= \rho \underset{\sim}{F} \bullet v + (\partial/\partial x_j)(T_{ji}v_i) - \nabla \bullet q + Q$$ (5)

where e is the internal energy, q is the diffusive heat flux, and Q is the heat supply. Kinetic energy production and advection can be eliminated by using (1) and (2) and the symmetry of the stress tensor:

$$\partial(\rho e)/\partial t + \nabla \bullet (v\rho e) = T_{ij}E_{ij} - \nabla \bullet \underset{\sim}{q} + Q,$$ (6)

where **E** is the *rate of deformation*, whose components are given by

$$E_{ij} = \tfrac{1}{2}(\partial v_i/\partial x_j + \partial v_j/\partial x_i) .$$ (7)

The first term on the right side of (6) is the *stress power*, the rate of interconversion between mechanical and thermal energy.

The concentration c of a trace solute is also governed by a balance law, and the corresponding flux form is

$$\partial c/\partial t + \nabla \bullet (vc) = - \nabla \bullet q_c,$$ (8)

where q_c is the diffusive flux of solute.

3.2 Constitutive Relationships

It is reasonable to treat crystal growth nutrients as Newtonian viscous fluids, so that the stress is given by

$$T = - p\mathbf{I} + \lambda(tr\mathbf{E})\mathbf{I} + 2\mu\mathbf{E} ,$$ (9)

where μ, λ, and p are material coefficients which may depend on density and temperature. Note that I placed the pressure p on the same footing as the shear viscosity μ and the second coefficient of viscosity λ. From (9), -p **I** is the stress at equilibrium, and hence determined by the thermal equation of state. If, as is often done, the nutrient is treated as an incompressible fluid, the matter is quite different. This will be discussed in the next section.

It is also reasonable to assume that the internal energy of the nutrient is proportional to the temperature T, i.e.,

$$e = c_v T, \tag{10}$$

and that heat diffusion is governed by Fourier's law

$$\mathbf{q} = - k \nabla T. \tag{11}$$

Diffusion of dilute solutes is usually taken to follow Fick's law

$$\mathbf{q}_c = - k_c \nabla c. \tag{12}$$

The case of general mixtures is more complicated, requiring concentration dependent constitutive equations and perhaps provision for chemical reactions between components.

3.3 The Constraint of Incompressibility and the Boussinesq Approximation

Crystal growth melts are liquids and hence nearly incompressible. It is therefore a reasonable approximation, and a convenient one as well, to assume that the fluid density is constant. However, this changes the theory in a fundamental way, and the changes have computational import.

Equation (1) is a "prognostic" equation, i.e., it provides an expression for the time-rate of change of the fluid density as a function of the instantaneous state of the flow field. As such, it could be used in a finite difference scheme, to advance the density field from one time step to the next. However, if the density is constant, (1) reduces to

$$\nabla \bullet \mathbf{v} = 0. \tag{13}$$

This is "diagnostic" rather than prognostic, i.e., it must be satisfied by the velocity field at every instant of time, but cannot be used to advance a flow variable to the next time step. Physically, the assumption of incompressibility is not a balance law but a *constraint*, and (13) is the equation of constraint.

Incompressibility also introduces some conceptual changes into the Newtonian constitutive relation (9). First, when (13) applies, tr $\mathbf{E}$ vanishes. The λ-term then drops out of (9), which is one of the advantages of assuming incompressibility.

More subtly, the nature of the pressure term changes completely. As stated above, the pressure in the compressible case is a function of density and temperature, as specified by the thermal equation of state. If that concept were taken without change in the constant-density case, it would imply that the pressure is a function of temperature alone. That, in turn, would imply that the fluid cannot be in static equilibrium at uniform temperature when a body force is present, which contradicts intuition.

The dilemma is resolved by recalling that constraints in mechanics are accompanied by force systems which are the reactions to the constraints. The reaction to the constraint of incompressibility is a pressure field which is not determined by a thermal equation of state. Rather, it is determined as part of the computational problem: at each time step, the pressure field must be such that it exerts internal forces on the fluid in such a way that the velocity field obeys (13).

This sounds like a lot of bother, but it isn't too difficult in practice. Without the assumption of incompressibility, it would be necessary to keep track of disturbances propagating through the liquid at the speed of sound, and that would be a much heavier price to pay.

There is one aspect of crystal growth flows, however, that forces us to reconsider the simplification of incompressibility. We know that thermal or solutal gradients in a liquid can lead to significant buoyant convection, even if the resulting density differences are only a few hundredths of a percent. The way out is to use the *Boussinesq approximation*: fluid density changes are ignored except in the body force term. This has been found to be widely useful in practice. The logical inconsistency of accounting for density changes "here" but not "there" can be made more palatable by taking a slightly different view: we can regard the liquid nutrient as an incompressible fluid acted upon by a body force which depends on the temperature and perhaps on the concentration of a solute.

With (7) and (9), the stress power can be expanded in terms of the velocity components and the pressure:

$$T_{ij}E_{ij} = -p\nabla \bullet \mathbf{v} + \lambda(\nabla \bullet \mathbf{v})^2 + \mu(\partial v_i/\partial x_j)(\partial v_i/\partial x_j + \partial v_j/\partial x_i) \, . \tag{14}$$

The first term on the right side represents heating by adiabatic compression; it vanishes when the fluid is incompressible. The rest of the right side comes from viscous dissipation. The λ term again vanishes for an incompressible fluid, but the μ term remains.

In the flow of crystal growth melts, however, heating by viscous dissipation is negligible compared with the external heating required to keep the nutrient molten. With this observation, plus incompressibility, the stress power drops out of (6), i.e., thermal energy is separately conserved. Equation (6) then simply relates the production, advection, and diffusion of heat.

3.4 The Magnetic Body Force

Some important crystal growth materials are excellent electrical conductors in their liquid state. This opens the possibility of using hydromagnetic effects to influence the fluid motion. The general study of magnetohydrodynamics begins with Maxwell's equations and Ohm's law for a medium in motion, simplified by the *quasi-static assumption*. That is, displacement currents and the effect of variations in the charge density are ignored.

In the general case, a formidable theory remains, but the study of hydromagnetic crystal growth flows is greatly simplified by the *liquid metal approximation*, that the decay time of electric currents and magnetic fields is very brief compared with the characteristic times of the fluid flow. This leads to a twofold simplification. First, the induced magnetic field is negligible, so that the magnetic induction $\mathbf{B}$ is that of the applied field. Second, if the applied field is steady, the electric field is irrotational and hence can be derived from a potential ϕ. Ohm's law then determines the electric current $\mathbf{j}$ in terms of ϕ according to

$$\mathbf{j} = \sigma(-\nabla\phi + \mathbf{v} \times \mathbf{B}) \, , \tag{15}$$

where σ is the electrical conductivity. An equation for ϕ is then obtained by substituting this into the conservation of charge relation

$$\nabla \cdot \mathbf{j} = 0 \,. \tag{16}$$

In flows depending on only two space dimensions, it is sometimes alternately possible to eliminate ϕ from the problem altogether, dealing instead with a *current function*, whose derivatives are proportional to the components of $\mathbf{j}$.

The influence of magnetic effects on the fluid flow is represented by the inclusion of a $\mathbf{j} \times \mathbf{B}$ body force in (3). In principle, the current $\mathbf{j}$ produces Joule heating, which contributes to the heat supply Q, but in practice this is usually negligible compared with external heat supplied at the boundaries.

3.5 Rotationally Symmetric Flow

When the flow has an axis of symmetry, it is more easily investigated in cylindrical coordinates (r, θ, z). In formulating the equations, we shall assume that all transport coefficients are constant. Since they are, in fact, temperature dependent, this is an approximation which holds only if temperature differences within the nutrient are not too severe.

An axial magnetic field can be applied without destroying the rotational symmetry. If the field acts in the increasing z-direction and has strength B, the equations governing the radial velocity u and the axial velocity w, derived from the momentum equation (3), are

$$
\frac{\partial u}{\partial t} + \frac{1}{r}\frac{\partial}{\partial r}\left(ru^2\right) + \frac{\partial}{\partial z}\left(wu\right) - \frac{v^2}{r}
$$
$$
= \frac{1}{\rho}\left(Bj_\theta - \frac{\partial p}{\partial r}\right) + v\left(\nabla^2 u - \frac{u}{r^2}\right) , \tag{17}
$$

$$
\frac{\partial w}{\partial t} + \frac{1}{r}\frac{\partial}{\partial r}\left(ruw\right) + \frac{\partial}{\partial z}\left(w^2\right)
$$
$$
= -\frac{1}{\rho}\frac{\partial}{\partial z}\left(p + \rho gz\right) + \alpha g(T - T_s) + v\nabla^2 w , \tag{18}
$$

where v is the azimuthal velocity, v is the kinematic viscosity, α is the volumetric expansion coefficient, and T_s is the solidification temperature.

The pressure can be eliminated from these equations by cross-differentiation and subtraction. This leads to a prognostic equation for the *vorticity* ω, defined by

$$\omega = \partial w/\partial r - \partial u/\partial z \,. \tag{19}$$

For numerically stable modeling of physically unstable flows, however, it is better to use ω/r rather than ω as a dependent variable. As ω/r plays a central role in a theorem due to A. V. Svanberg dealing with vorticity transfer in rotationally symmetric motion, I have suggested that it be termed the *Svanberg vorticity*, denoted by the symbol S. Combining (17) and (18) leads to

$$\frac{\partial S}{\partial t} + \frac{1}{r}\frac{\partial}{\partial r}(ruS) + \frac{\partial}{\partial z}(wS) + \frac{\partial}{\partial z}\left(\frac{v^2}{r^2}\right)$$
$$= \frac{\alpha g}{r}\frac{\partial T}{\partial r} - \frac{B}{\rho r}\frac{\partial j_\theta}{\partial z} + \frac{v}{r}\frac{\partial}{\partial r}\left[\frac{1}{r}\frac{\partial}{\partial r}(r^2 S)\right] + v\frac{\partial^2 S}{\partial z^2} \ . \tag{20}$$

The incompressibility condition (13) is automatically satisfied by expressing the meridional velocity components in terms of a Stokes streamfunction ψ according to

$$u = \frac{1}{r}\frac{\partial\psi}{\partial z} \ , \quad w = -\frac{1}{r}\frac{\partial\psi}{\partial r} \ . \tag{21}$$

Using this *vorticity-streamfunction formulation* for the meridional flow simplifies computation considerably. However, it does not eliminate the need to solve a diagnostic equation necessitated by the constraint of incompressibility: combining (21) with the definition of S leads to

$$\frac{\partial}{\partial r}\left(\frac{1}{r}\frac{\partial\psi}{\partial r}\right) + \frac{1}{r}\frac{\partial^2\psi}{\partial z^2} = -rS \ . \tag{22}$$

The momentum equation (3) also leads to an equation for the azimuthal velocity v, but it is better not to use it, because $\rho r v$ rather than ρv is the physically conserved quantity (angular momentum per unit volume). Thus, false production of angular momentum is more easily avoided if we introduce the *swirl* $\Omega = rv$, which is governed by the prognostic equation

$$\frac{\partial\Omega}{\partial t} + \frac{1}{r}\frac{\partial}{\partial r}(ru\Omega) + \frac{\partial}{\partial z}(w\Omega)$$
$$= -\frac{B}{\rho}rj_r + \frac{v}{r}\frac{\partial}{\partial r}\left[r^3\frac{\partial}{\partial r}\left(\frac{\Omega}{r^2}\right)\right] + v\frac{\partial^2\Omega}{\partial z^2} \ . \tag{23}$$

Ohm's law (15) yields equations for the three components of electric current. Because of the rotational symmetry, the θ derivative of the potential vanishes. Therefore

$$j_r = \sigma(-\partial\phi/\partial r + B\Omega/r), \tag{24}$$

$$j_\theta = -\sigma Bu, \tag{25}$$

$$j_z = -\sigma\partial\phi/\partial z. \tag{26}$$

With (21) and (25), the magnetomotive term in (20) can be written in terms of the streamfunction:

$$-(B/\rho r)\partial j_\theta/\partial z = (\zeta/r^2)\partial^2\psi/\partial z^2, \tag{27}$$

with $\zeta = \sigma B^2/\rho$.

The conservation of charge relation (16) is automatically satisfied if we express the meridional current in terms of a *current function* ψ_J, such that

$$j_r = \frac{\sigma B}{r}\frac{\partial\psi_J}{\partial z}, \quad j_z = -\frac{\sigma B}{r}\frac{\partial\psi_J}{\partial r}. \tag{28}$$

The magnetomotive term in (23) can then be written in terms of the current function:

$$-(Brj_r/\rho) = -\zeta\partial\psi_J/\partial z. \tag{29}$$

Substituting (28) into (24) and (26), then cross-differentiating and subtracting to eliminate ϕ, yields a diagnostic equation for the current function:

$$\frac{\partial}{\partial r}\left(\frac{1}{r}\frac{\partial\psi_J}{\partial r}\right) + \frac{1}{r}\frac{\partial^2\psi_J}{\partial z^2} = \frac{1}{r}\frac{\partial\Omega}{\partial z}. \tag{30}$$

With viscous dissipation and Joule heating neglected, transfer of heat is governed by

$$\frac{\partial T}{\partial t} + \frac{1}{r}\frac{\partial}{\partial r}(ruT) + \frac{\partial}{\partial z}(wT) = \kappa\nabla^2 T, \tag{31}$$

where $\kappa = k/(\rho c_v)$. The transport of a dilute solute follows

$$\frac{\partial c}{\partial t} + \frac{1}{r}\frac{\partial}{\partial r}(ruc) + \frac{\partial}{\partial z}(wc) = k_c\nabla^2 c. \tag{32}$$

Equations (20), (22), (23), (30), (31), and (32), comprise a differential equation system for examining crystal growth flows under a wide range of conditions. To tie them down to Czochralski flow, an appropriate set of boundary conditions must be specified.

3.6 Boundary Conditions for Czochralski Flow

The crucible wall and bottom and the growth interface are no-slip surfaces. Hence the value of the swirl Ω is directly specified on each point of them. There are also straightforward procedures for translating no-slip into conditions on the vorticity.

Strictly speaking, the free surface of the melt should be treated as a movable boundary. However, Czochralski nutrients are molten metals or other heavy liquids, and the forces acting are relatively gentle. Hence the free surface tends to remain nearly flat and stationary. Taking the upper boundary of the melt to be a fixed, flat surface is called the *bulk flow approximation*, and it is widely used. Since no significant externally applied stresses act on the free surface, velocity conditions on this part of the boundary are determined entirely by thermocapillarity: the normal derivative of the radial velocity is proportional to the radial derivative of temperature, and the axial velocity vanishes (bulk flow approximation); because of rotational symmetry, thermocapillarity does not directly drive the azimuthal flow, so that the normal derivative of the swirl vanishes on the free surface.

In their full generality, thermal boundary conditions on Czochralski melts can be quite complicated. Radiative heat transfer between the melt and its surroundings (crucible, heater, the solid crystal, etc.) must be accounted for in some contexts. For melt convection studies, however, it is often sufficient to use simpler conditions, provided the overall heat balance approximates that which occurs in practice. For example, the temperature on the crucible and on the growth interface may be specified, along with a relatively simple model for radiation from the free surface.

For growth in a magnetic field, electrical boundary conditions must also be provided. If the crucible is made of a non-conducting material, such as silicon dioxide, the crucible and the free surface are insulating boundaries. Electrical conditions at the growth interface depend on the conductivity of the solid crystal.

Boundary conditions on the concentration of a dilute solute depend on the nature of the solute. For deliberately added dopants, the only source-sink mechanism is incorpo-

ration into the growing crystal. For dissolved gasses, on the other hand, there is evaporation loss at the free surface, and ablation of the crucible may act as a source.

Symmetry conditions apply at r = 0. Thus all fluxes of momentum, heat, solute, and
electric current are in the axial direction. Subsidiary conditions on the swirl and Svanberg
vorticity at the axis of symmetry are not obvious, but can be derived by requiring that the
flow have no singularities at the axis.

4. A SURVEY OF DEVELOPMENTS IN THE MODELING OF CZOCHRALSKI FLOW

The very earliest use of digital simulation to investigate Czochralski flow was by
Kobayashi and Arizumi in 1970 [5]. Their model assumed steady flow, which is valid for
small crucibles. It was able to reproduce flow patterns which resembled those found experimentally in flow visualization studies. An important result found with this model is that
the nature of Czochralski flow can depend qualitatively upon the Grashof number, a
dimensionless parameter which measures the relative importance of buoyancy and
centrifugal pumping. At low values, the flow is primarily a "tip vortex" generated by the
rotating crystal. Above a critical value, buoyancy generates a significant counter-flow. At
sufficiently high Grashof number, it can completely overwhelm the centrifugal flow generated by the rotating crystal. This offered some insight about why certain matters of interest
to the crystal grower, such as interface shape and rate of bubble inclusion, can change
qualitatively, and sometimes abruptly, when the flow parameters are modified.

For crucibles in the size range commonly used in practice, simulation requires the
time-dependent equations. The flow field often fluctuates indefinitely, never settling down
to a completely steady state. This can be so even if the heating and the rotation are absolutely steady.

Nor is the time variation limited to small-scale fluctuations. In [6], I attempted to
determine the steady-state flow field for silicon melt, using parameters that were appropriate for industrial-scale growth in the mid-1970's. The importance of thermocapillary
convection was not appreciated at the time, and it was not incorporated into the computation. Unexpectedly, even the qualitative appearance of the flow field failed to reach a

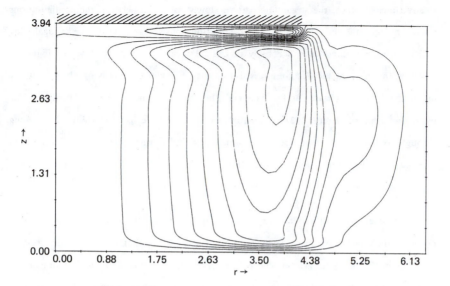

Fig. 2. Simulated streamlines in the meridional plane at a time when the buoyant cell fills most of the crucible.

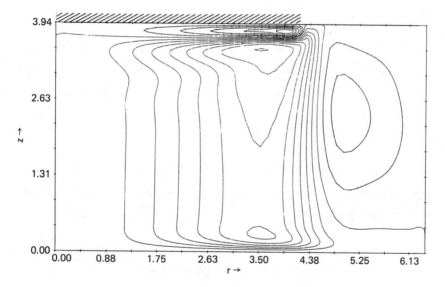

Fig. 3. Streamline pattern about one simulated second after that of Fig. 2. The buoyant cell forms a Taylor column beneath the growth interface.

steady state. Instead, the streamline pattern slowly oscillated between the configurations shown in Figures 2 and 3. In Figure 2, the motion consists entirely of a centrifugal eddy under the growth interface and a buoyant cell filling most of the crucible. In Figure 3, however, the buoyant cell has compressed itself into a Taylor column lying almost entirely under the growth interface, and a back-eddy has formed beneath the free surface. The oscillation repeated every 2 1/3 simulated seconds, and persisted without diminishing through six cycles. At that point I became convinced it was a permanent feature, and stopped the computation. Similar oscillations were found by others, giving persuasive evidence that striations in crystals are hydrodynamic in origin.

The time-dependent model was used to take a detailed look at the transition that occurs as the Grashof number is increased [8]. The work was motivated by a conversation with Phillip Yin, a crystal grower who had observed that the flow in a gadolinium-gallium-garnet melt was quite different from that in neodymium-gallium-garnet. The two compounds are closely related, except that the latter has a much lower volumetric expansion coefficient.

The study was carried out for a non-rotating crucible, so there was no tendency for Taylor columns to form. The streamline pattern for very low buoyancy is shown in Figure 4. The buoyant cell is completely encapsulated by the tip vortex, and hence is isolated from the growth interface. At the opposite extreme, Figure 5, the buoyant cell is dominant, and is in fact strong enough to generate two back-eddies.

With a time-dependent model, the transition range itself can be examined. Figure 6 shows a streamline pattern within this range. The various eddies interact with one another, and the flow is decidedly unsteady.

Since violent coarse-grained motion within the nutrient should correlate with severe fluctuations of heat and solutes to the growth interface, crystal growers have sought ways to calm the flow. Some nutrients, molten silicon for example, are excellent conductors of electricity. This suggested the possibility of growing crystals in a large magnetic field, so that the fluid motion would be damped by induction drag. Many investigations resulted, and the work has recently been reviewed [8].

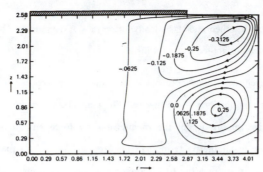

Fig. 4. Streamline pattern when the flow is dominated by the tip vortex.

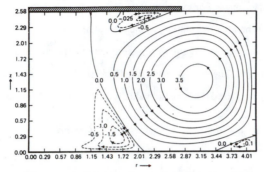

Fig. 5. Streamline pattern when the flow is dominated by the buoyant cell.

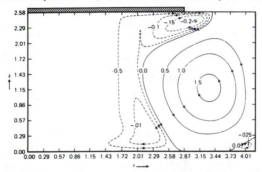

Fig. 6. Unstable streamline pattern in the transition regime.

In the case of Czochralski growth, a natural choice for the magnetic field is a uniform field acting parallel to the axis of the crystal, since it does not destroy the axial symmetry of the configuration. The crystal grower is motivated to preserve symmetry because of his interest in crystal uniformity, but a by-product is that investigation of the flow remains more accessible to the theoreticial hydrodynamicist. Axisymmetric computer models remain valid, and have in fact led to useful insight.

Also, fields large enough to produce dramatic damping are frequently large enough to permit asymptotic simplifications of the theory. The magnetic interaction parameter, which measures the relative importance of the magnetomotive force and fluid inertia, is large enough that the non-linear inertia terms can be dropped from the equations of motion. This linearization makes it possible to obtain complicated solutions to the equations by superimposing simpler ones. For example, flow generated by buoyancy alone can be superposed on the flow generated by rotation of the crystal and crucible. Another simplification which arises when the field is large is that viscous effects are confined to boundary layers and inner shear layers. The flow domain can therefore be divided into subregions where asymptotic solutions can be developed individually and then matched together.

The computational and asymptotic approaches to Czochralski flow in a uniform axial field complement each other well. The former, unlike the latter, is not limited to cases where the field is strong. On the other hand, the asymptotic approach is at its best for large fields, where the computational method is hampered by the need for an exceedingly fine grid to resolve solute transport through boundary layers. There is a region of overlap where both approaches work well. For commercial Czochralski growth of silicon, this is about 0.2 to 0.3 tesla. Simulations in this range have shown excellent agreement between the methods.

Unfortunately, using a uniform axial field to calm Czochralski flow entails a drawback. The finished crystals show an unacceptably large radial variation of solute concentration. For gallium-doped silicon grown in a 0.3 tesla axial field, the gallium concentration in the crystal drops off by a decade and a half from axis to edge.

Examining the simulated flow patterns and solute distributions in the melt explained the phenomenon. The segregation coefficient for gallium in silicon is very low, specifically, 0.008. That is, only 0.8% of the gallium reaching the growth interface is incorporated into the growing crystal; the rest is rejected back into the melt. In a well-mixed melt, this is quickly redistributed and the concentration isopleths remain nearly horizontal. In a strong axial field, however, the radial flow beneath the crystal is severely damped since it crosses lines of force at right angles. The excess gallium is swept away only by a shear layer found near the crystal edge. Migration of gallium toward the edge takes place almost entirely through the slow process of diffusion, so that a radial gradient of gallium concentration is set up. Before equilibrium between rejection and diffusion is established, the concentration is much higher at the axis than at the edge.

The case of gallium in silicon is exceptionally striking because of the low segregation coefficient, but the difficulty occurs in other systems as well. Simulations for boron in silicon indicated that, for growth in a 0.2 tesla axial field with a crystal rotation rate of 16 RPM, the boron concentration near the crystal edge is about half that at the axis. The case of volatile solutes is more complicated because evaporation from the melt free surface forces a very low concentration in edge regions. Thus, even though the segregation coefficient for oxygen in silicon is somewhat greater than unity, simulation [4] and experiment both show a marked drop-off in oxygen concentration from axis to edge for growth in a 0.1 tesla axial field.

Thus the very mechanism that eliminates chaotic flow in the bulk of the melt also eliminates advective redistribution of solute in the neighborhood of the growth interface, and diffusive redistribution is too weak to do a satisfactory job. Advective transport is not always undesirable.

Disappointment with uniform axial fields has led to consideration of other field configurations. A transverse magnetic fields offers the advantage that the lines of force are parallel to the crystal face. Thus advective redistribution of rejected solvent is relatively uninhibited along those radial directions that cross lines of force at small angles. However, a transverse field destroys the axial symmetry. A point on the crystal face sees fluctuations at twice the crystal rotation rate. Thus a strong transverse field can be expected to produce non-uniformities in the crystal with an axial scale equal to the pull velocity times half the crystal rotation period.

A possible solution is to apply a transverse field which is strong enough to suppress fluctuations, but weak enough to produce only small deviations from axisymmetry. There is theoretical evidence for a range of field strengths satisfying both requirements. Such evidence is hard to come by because the absence of axial symmetry is a major complication. Some work, however, has been done. Direct three-dimensional simulation is one approach [9]. Another is to use asymptotic methods in concert with computation. The latter approach, reviewed in [8], was the method that provided evidence for a range of fields strengths strong enough to suppress fluctuations but weak enough to preserve most of the axial symmetry.

An alternate way to get the advantages of magnetic damping without the disadvantages is to use an appropriately shaped non-uniform axisymmetric field. This approach was investigated in [10], which presented the first digital simulations for a non-uniform magnetic field. The field was produced by two coils carrying equal and opposite DC electric currents with symmetry about the plane of the free surface. This field is essentially radial in the diffusion layer under the growth interface. The centrifugal flow generated by crystal rotation is therefore along the lines of force and is not subjected to magnetic damping.

In [8], additional characteristics for a favorably shaped field were discussed. One way that a magnetic field can help control oxygen inclusion in a silicon crystal is to influence the incorporation of oxygen into the melt. The mechanism for this is ablation of the silica crucible, which is exacerbated by advective transport of oxygen away from the crucible wall and bottom into the bulk of the melt. For reducing transport near the wall, a uniform axial field is no help at all. The lines of force are parallel to the wall, so the magnetomotive force does not oppose the vertically upward buoyancy force. Instead, buoyancy drives a relatively high-velocity jet, which scours oxygen from the crucible. What is more desirable is a magnetic field with a significant normal component over both the vertical wall and the crucible bottom. This is not hard to arrange: the field investigated in [10] has this characteristic. Since the criterion should be met at all melt depths during the growth process, the optimally tailored magnetic field will be time-dependent.

Meaningful simulation of Czochralski flow in a shaped axisymmetric field is not as difficult as transverse field studies. The magnetic field does not destroy the axial symmetry. In fact, it should preserve it in many cases where, because of instability, non-magnetic

Czochralski flow would be unsymmetric. The governing equations involve a minor modification of those for a uniform axial field, so that much of the existing numerical methodology carries over: the modeler can focus his efforts on good physics and high resolution.

5. OUTLOOK

The importance of hydrodynamics in crystal growth, and the feasibility of doing it, is now well established. Difficult problems, such as incorporation of turbulence, still remain. However, the groundwork has been laid, and the field is very active.

Perhaps the most important work to be carried out during the next decade is analysis of heat and mass transfer in the entire crystal growth system. Such studies are already underway, and have been recently reviewed [11]. From the viewpoint of the crystal grower, hydrodynamics is already a tool for understanding. Investigations of the complete system — melt, furnace, solid crystal, and ambient gas — have the potential to make it a design tool as well.

References

1.   *Energy Optimisation in Manufacturing and Materials Processing,* edited by R. K. Shah, H. Md. Roshan, V. M. K. Sastri, and K. A. Pamanabhan. (Indian Institute of Technology, Madras, 1989).

2.   W. E. Langlois, in *Annual Review of Fluid Mechanics,* edited by Milton Van Dyke, J. V. Wehausen, and John L. Lumley. (Annual Reviews, Palo Alto, 1985), **17**, p. 191.

3.   K. M. Kim and W. E. Langlois, J. Elechem Soc. **136**, p. 1137 (1989).

4.   K. M. Kim and W. E. Langlois, *Proc. 6th Int. Symp. on Silicon Material Sci. and Technology,* edited by H. R. Huff, K. G. Barraclough, and J. Chikawa. (The Electrochemical Society, Inc., Pennington N. J., 1990), p. 81.

5.   N. Kobayashi and T. Arizumi, Jpn. J. Appl. Phys. **9**, p. 361 (1970).

6. W. E. Langlois, J. Cryst. Growth **42,** p. 386 (1977).

7. W. E. Langlois, J. Cryst. Growth **46,** p. 743 (1979).

8. W. E. Langlois, K. M. Kim, and J. S. Walker, in *Fluids Engineering Seminar: Korea-U.S. Progress,* edited by Jong Hyun Kim, Jae Min Hyun, and Chung-Oh Lee. (Hemisphere, Washington, 1991), p. 551.

9. M. Mihelčič and K. Wingerath, J. Cryst. Growth **82,** p. 318 (1987).

10. T. W. Hicks, A. E. Organ, and N. Riley, J. Cryst. Growth **94,** p. 213 (1989).

11. R. A. Brown, T. A. Kinney, P. A. Sackinger, and D. E. Bornside, J. Cryst. Growth **97,** p. 97 (1989).

Modeling Complex Phenomena in Fluids

A. Garcia, D. Morris, J. Stroh and C. Penland

At the macroscopic level, fluids have traditionally been modeled using numerical schemes based on partial differential equations (e.g. Navier-Stokes equations). In recent years there has been a growing interest in the simulation of complex flows using particle methods (e.g. Molecular Dynamics and Cellular Automata). In this paper a variety of projects currently underway at San Jose State University are described. The unifying theme is the application of unconventional methods to fluid mechanics problems.

1. INTRODUCTION

In the past 40 years, computers have revolutionized the way we study fluids. Some of the earliest computational work in physics involved the calculation of transport coefficients by Molecular Dynamics.[1] Many types of fluid flow in systems ranging from stellar interiors to the chambers of the human heart have been modeled by partial differential equations. The Navier Stokes equations (or their reduced variants) may be solved numerically by finite difference, finite element or spectral techniques.[2]

In more recent years there has appeared a full spectrum of new algorithms in computational fluid dynamics (see Table 1). These new schemes typically fill an "ecological niche" in which they are superior to more

A. Garcia, D. Morris and J. Stroh: Department of Physics, San Jose State University, San Jose CA 95192-0106

C. Penland: CIRES, University of Colorado, Boulder CO 80309-0216

conventional methods. It is also common to find simulations which are hybrids, combining several methods. For example, the general circulation models used in weather prediction are fundamentally partial differential equation solvers but some elements, such as the air-sea interaction, are included in a phenomenological fashion.

Table 1. Spectrum of computational methods used in fluid dynamics.

Computational Method	Main Fields
Molecular Dynamics	Chemistry, Statistical Physics
Direct Simulation Monte Carlo	Rarefied Gas Dynamics
Lattice Boltzmann	Special Applications
Cellular Automata	Special Applications, (e.g. Flow through porous media)
Super-particle methods	Plasma Physics, Astrophysics
Navier-Stokes Integrators (CFD)	General Applications
Phenomenological Models	Complex systems

Despite the diverse nature of available algorithms most physicists only use the one or two which are well-known in their field of work. In this paper we briefly review three projects which have been underway at San Jose State in the past year. Each project involves using a computational method for a problem which is very different from the field in which that method is commonly used. We hope that the work presented here encourages a more open-minded approach in the use of these algorithms.

2. MOLECULAR DYNAMICS

Molecular dynamics (MD) simulations model a fluid by computing the trajectory in classical phase space. These simulations have traditionally been used to compute fluid properties such as the equation of state. In the past few years there have been several MD simulations of fluid flow.[3]

BHATTACHARYA and LEE used an MD simulation to study the velocity profile in Poiseuille flow.[4] Near the wall, the fluid's velocity does not go to zero even if the wall is perfectly thermalizing. Extrapolating the velocity profile one finds that the velocity would go to zero at a distance σ inside the wall (see Fig. 1). According to Maxwell's theory this distance should be

$$\sigma = \alpha \lambda \tag{1}$$

where λ is the mean free path and α (≈ 1.1) is the slip coefficient. We define the dimensionless slip length as,

$$\text{slip length} \equiv \frac{\sigma}{L} = \frac{\alpha\lambda}{L} = \alpha \, Kn \tag{2}$$

where L is the characteristic length (channel width) and Kn ($= \lambda/L$) is the Knudsen number. Thus according to Maxwell's theory the dimensionless slip length should be proportional to the Knudsen number.

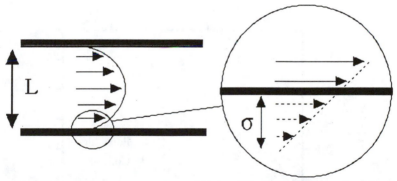

Figure 1: Schematic representation of Poiseuille flow. Notice that the flow velocity is non-zero at the wall; extrapolating we find that it goes to zero at a distance σ inside the wall.

Figure 2 shows BHATTACHARYA and LEE's results for Poiseuille flow along with our results for planar Couette flow. Notice that for large Knudsen number the slip length deviates from the Maxwell theory. BHATTACHARYA and LEE proposed that the slip length might be proportional to log(Kn) but gave no physical justification for their proposal.

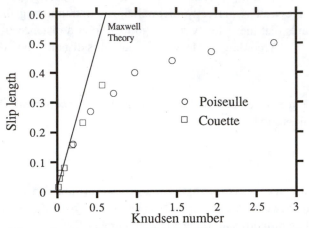

Figure 2: Dimensionless slip length versus Knudsen number for Couette and Poiseuille flow. The Poiseuille data is from [4]; the Couette data is from [6].

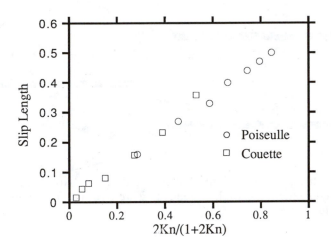

Figure 3: Dimensionless slip length versus 2Kn/(1+2Kn) for Couette and Poiseuille flow (see (3)). Compare with Fig. 2.

The results in Fig. 2 are very reminiscent of the Knudsen number dependance of the shear stress at the wall in a rarefied flow. According to Lees theory [5], the kinematic shear stress, τ, in Couette flow varies with Knudsen number as

$$\tau \propto \frac{2\,Kn}{1 + 2\,Kn} \tag{3}$$

Figure 3 shows that for both Couette and Poiseuille flow the slip length appears to have this same dependance on Knudsen number. For more details on these results see reference [6].

3. DIRECT SIMULATION MONTE CARLO

Rayleigh-Bénard convection is a paradigm instability; at a critical Rayleigh number there is a bifurcation between states of purely conductive heat flow and buoyancy-driven convection.[7] The nature of the hydrodynamic fluctuations near this transition point has been studied theoretically by a variety of methods.[8] Several careful experiments have measured the variation in the heat flux near the onset of convection.[9] However, quantitative comparison between fluctuating hydrodynamics calculations and laboratory experiments reveals significant discrepancies.[10]

The Direct Simulation Monte Carlo (DSMC) method was developed by G. A. Bird for the study of rarefied gas flows.[11] It has been described as "The dominant predictive tool in rarefied gas dynamics for the past decade...".[12] The method has also been used successfully to study fluctuations in simple nonequilibrium systems.[13]

The DSMC method was used to simulate a dilute gas of 50,000 hard sphere particles in a Rayleigh-Bénard configuration. The system was studied near the onset of convection, i.e. about the critical Rayleigh number. Principal Oscillation Pattern (POP) analysis [14] was used to measure the fluctuating hydrodynamic noise. The spatial correlation of the noise was found to be very similar to its equilibrium form. From fluctuating hydrodynamics, the noise variances for the x-velocity, Q^{Vx}, y-velocity, Q^{Vy}, and temperature, Q^T, depend on local temperature, T, as,

$$Q^{Vx} \text{ or } Q^{Vy} \propto T^{3/2} \qquad \text{and} \qquad Q^T \propto T^{5/2}. \tag{4}$$

Using the data from the DSMC simulation for a system near the critical point, the POP analysis gives

$$Q^{v_x} \propto T^{1.19} \qquad Q^{v_y} \propto T^{1.35} \qquad Q^T \propto T^{2.35}. \tag{5}$$

These results are illustrated in Fig. 4. This shows that the noise terms in the Landau-Lifshitz fluctuating hydrodynamics retain their equilibrium form near the critical Rayleigh number. These results are described in more detail in [15].

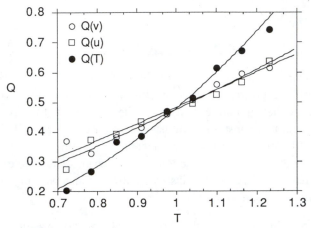

Figure 4: Noise variance as a function of temperature in a system near the critical Rayleigh number. The open squares and circles are Q^{v_x} and Q^{v_y}, respectively; the solid circles are Q^T. Solid lines are the least square power law fits to the data.

4. CELLULAR AUTOMATA

Interesting fractal aggregates have been found in electrochemical deposition studies.[16] Several theoretical models, for example the dielectric breakdown model and various random walker models, have been studied by the San Jose group.[17] An important element missing from these simplified models is the inclusion of hydrodynamics. Towards this end, we have introduced a multi-species lattice gas model.

There are three types of particles in our model: the static aggregate particles, the mobile ions and the mobile neutrals. The dynamics of the mobile particles is similar to that of the HPP model.[18] When a mobile ion encounters an aggregate particle it "sticks" to the aggregate (the ion is turned into an aggregate particle). The lattice gas interaction of ions and neutrals incorporates simple hydrodynamic effects. The model was implemented using a CAM-PC board on an AT clone.[19] Using the CAM-PC board the model runs very quickly; a 10,000 particle aggregate takes only seconds to form.

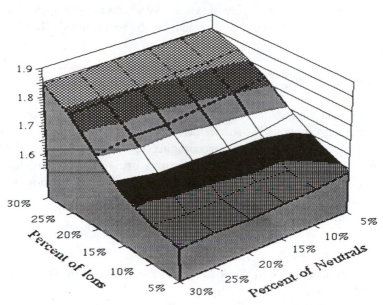

Figure 5: Box dimension on the vertical axis as a function of initial percentage of ions and percentage of neutrals.

One of the most frequently studied properties of aggregate clusters is their fractal dimension.[20] The box (capacity) dimension as a function of ion and neutral density is shown in fig. 5. Interestingly, the fractal dimension depends weakly on the concentration of neutrals. This is not an intuitively obvious result since the mean free path of the ions depends strongly on the total density of particles. For more details on these simulations see [21].

5. FUTURE WORK

One of the outstanding problems in computational fluid mechanics is the simulation of high Reynolds number flows, i.e. turbulence. For engineering purposes, phenomenological turbulence models (e.g. Reynolds stress models) are commonly used. To simulate a turbulent flow at the level of the Navier-Stokes equations is a difficult problem due to numerical instabilities. Furthermore, one must simulate a three dimensional system.

It is difficult to achieve high Reynolds number in particle simulations for the following reason: The kinematic viscosity in a gas may be approximated as $\eta \approx \lambda c$ where λ is the mean free path and c is the sound speed. The Reynolds number is then

$$Re = \frac{u\,L}{\eta} \approx \frac{u}{c}\frac{L}{\lambda} \tag{6}$$

where L and u are the characteristic length scale and flow velocity. Assume that our system contains N particles per cubic mean free path. A back of the envelope calculation yields a discouraging result: to achieve a Reynolds number of 10^4 in a sub-sonic flow requires using $N \cdot 10^{12}$ particles. Our group, in collaboration with NASA Ames and Lawrence Livermore National Lab, is looking into devising new methods for simulating three dimensional turbulent flow.

The authors would like to thank Lui Lam, Lar Hannon and Berni Alder for helpful discussions.

References
1. *Simulation of Liquids and Solids*, edited by G. Ciccotti, D. Frenkel and I. McDonald. (North-Holland, Amsterdam,1987).
2. D. Anderson, J. Tannehill and R. Pletcher, *Computational Fluid Mechanics and Heat Transfer* (Hemisphere Publ., New York, 1984).
3. *Microscopic Simulations of Complex Flows*, edited by M. Mareschal. (Plenum, New York, 1990).
4. D. Bhattacharya and G. Lee, Phys. Rev. Lett. **62**, 897 (1989); D. Bhattacharya and G. Lee, Phys. Rev. A **43**, 761 (1990).

5. W. Vincenti and C. Kruger, *Introduction to Physical Gas Dynamics* (R.E. Krieger Publ., Malabar FL, 1986).

6. D. Morris and A. Garcia, submitted to Phys. Rev. A (1991).

7. S. Chandrasekhar, *Hydrodynamic and Hydromagnetic Stability* (Dover Press, New York, 1981).

8. V.M. Zaitsev and M.I. Shliomis, Sov. Phys. JETP **32**, 866 (1971); H.N.W. Lekkerkerker and J.P. Boon, Phys. Rev. A **10**, 1355 (1974); T.R. Kirkpatrick and E.G.D. Cohen, J. Stat. Phys. **33**, 639 (1983); R. Schmitz and E.G.D. Cohen, J. Stat. Phys. **38**, 285 (1985).

9. G. Ahlers, M.C. Cross, P.C. Hohenberg and S. Safran, J. Fluid Mech. **110**, 297 (1981); R.P. Behringer and G. Ahlers, J. Fluid Mech. **125**, 219 (1982); C.W. Meyer, G. Ahlers and D.S. Cannell, Phys. Rev. Lett. **59**, 1577 (1987); G. Ahlers, C.Meyer and D. Cannell, J. Stat. Phys. **54**, 1121 (1989).

10. H. VanBeijeren and E.G.D. Cohen, Phys. Rev. Lett. **60**, 1208 (1988); ibid., J. Stat. Phys. **53**, 77 (1988).

11. G.A. Bird, *Molecular Gas Dynamics* (Claredon Press, Oxford, 1976).

12. E. Muntz, Ann. Rev. Fluid Mech. **21**, 387 (1989).

13. A. Garcia, Phys. Rev. A **34**, 1454 (1986); M. Malek Mansour, A. Garcia, G. Lie and E. Clementi, Phys. Rev. Lett. **58**, 874 (1987); A. Garcia, M. Malek Mansour, G. Lie and E. Clementi, J. Stat. Phys. **47**, 209 (1987).

14. C. Penland, Mon. Wea. Rev. **117**, 2165 (1989).

15. A. Garcia and C. Penland, J. Stat. Phys. **64**, 1121 (1991).

16. A. Witten and L.M. Sander, Phys. Rev. Lett., **47**, 1400 (1981).

17. L. Lam, R.D. Pochy and V.M. Castillo, in: *Nonlinear Structures in Physical Systems*, edited by L. Lam and H.C. Morris (Springer, New York, 1990); M.A. Guzman, R.D. Freimuth, P.U. Pendse, M.C. Veinott, and L. Lam, in: *Nonlinear Structures in Physical Systems*, edited by L. Lam and H.C. Morris (Springer, New York, 1990). R.D. Pochy, A. Garcia, R.D. Freimuth, V.M. Castillo and L. Lam, Physica D **51**, 539 (1991).

18. J. Hardy, O. De Pazzis, and Y. Pomeau, Phys. Rev. A13, 1949 (1976).

19. T. Toffoli and N. Margolus, *Cellular Automata Machines* (MIT Press, Cambridge, Massachusetts, 1987)

20. T.S. Parker and L.O. Chua, *Practical Numerical Algorithms for Chaotic Systems* (Springer, New York, 1989).

21. J. Stroh, SJSU Physics Dept. Technical Report, TR91-1 (1991).

Part VI
Complex Patterns

Consensus in Small and Large Audiences

V. Kanevsky, A. Garcia and V. Naroditsky

We consider the dynamics of a population in which individuals select a color from a finite set of colors. Each individual in the population has a "neighborhood" and the choices made by the neighbors influence the selection process of that individual. We are interested in the dynamics of the evolution of such a system to the monochromatic state of unison. We consider how the mean time to unison varies with the sizes of the population and the neighborhood, the geometry of the neighborhoods and the initial number of colors. Some analytical results are presented for the case where the size of the neighborhood is the entire population. Results of computer experiments for a variety of scenarios are presented and discussed.

1. INTRODUCTION

The phenomenon of self-organization exhibits itself in many dynamical systems both deterministic and stochastic. The difficulty of analytic investigation of the dynamics of such systems makes the solution of model problems much appreciated. The purpose of the present paper is to consider a model which attempts to explain the behavior of entities when the activity of each individual depends on the state of individuals in its neighborhood. One example is the synchronization in clapping in an auditorium (an ovation). The mathematical model of

V. Kanevsky, V. Naroditsky: Department of Mathematics and Computer Science, San Jose State University, San Jose CA 95192-0103

A. Garcia: Department of Physics, San Jose State University, San Jose CA 95192-0106

this phenomenon was motivated by the observations by one of us (VK) at communist party meetings during the "pre-glasnost" era. The model, however, is applicable to many different systems in the natural and social sciences.

Consider an audience in which each individual selects a color from a finite set of colors. Another way to view this is that each individual casts a vote given a finite set of choices. The colors or choices can be interpreted as distinct phases in the clapping of an audience; when all individuals clap with the same phase then it is synchronized. Each individual in the audience has a "neighborhood". The choices made by the members of this neighborhood influence the voting of that individual. All neighborhoods are congruent in a certain sense which follows from the context.

The dynamics of the system is as follows: initially each individual selects a color according to a given distribution. At each time step each member of the audience selects a new color (i.e. there are rounds of voting). The process is stochastic and an individual's vote is a function of the previous selections made by the members of his neighborhood. The system eventually reaches an absorbing state (the uniform, final state). We are interested in the dynamical evolution of such a system to the monochromatic state of unison; specifically: How does the mean time till absorption vary with the size of the audience? How does it depend on the size and geometry of the neighborhood or on the number of colors (choices)? Does the system display any critical phenomena?

Many variants of similar models appear in the literature. For example, in mathematical genetics we have the Fisher-Wright model.[1,2] In this model, in a finite population of size N, each member has one of two alleles (colors), A or B. The random variable is $x(t)$, the fraction of the population with allele A at time t. The dynamics of this system is as follows: the allele of each member of the population at time $t+\Delta t$ is A with probability $x(t)$ and B with probability $1-x(t)$. A related model is described by Mansour and De Palma.[3] In these models the neighborhood of a member of the population is the entire population; similar models with local neighborhoods have also been studied. [4]

In physics, our model is related to a class of time-dependent ferromagnet models.[5] There are many such models with the Ising model being the simplest. In these models a site may change its state (spin) with a probability given by the Boltzmann factor, $\exp(\Delta E/k_B T)$, where ΔE is the change in the interaction energy between the site and its neighbors, T is the temperature and k_B is Boltzmann's constant. In our model the probability that at the next step a site changes its state $\lambda \to \lambda'$ is proportional to the fraction of neighbors in state λ' on the previous time step.

2. ANALYTIC RESULTS

2.1 General Formulation

Let $A = \{a_1, ..., a_N\}$ be an arbitrary set of fixed elements (members of the audience). The neighborhoods U_a of each $a \in A$ are chosen such that the cardinality, $\text{card}(A \cap U_i) = n$, does not depend on i, i.e. the size of the neighborhood is the same for all members of the population. Each a_i takes one of the m colors from a set $\Lambda = \{\lambda_1, ...,\lambda_m\}$; $\lambda(a_i) \in \Lambda$ is the color of a_i.

The dynamics of the system is described by a Markov process. Each a_i assumes the new color λ_j on the next step $t+\Delta t$ with a probability proportional to the number of neighbors with this color on the current step t. The function $\lambda(a_i,t)$ is the color of a_i at time t. The dynamics of $\lambda(a_i,t)$ given by the conditional probability is

$$P(\lambda(a,t+\Delta t) = \lambda_j \mid \lambda(a_i,t), i=1,...,N) = \frac{\alpha(j,a,t)}{\text{card}(U_a)} = \frac{\alpha(j,a,t)}{n} \qquad (1)$$

where $\alpha(j,a,t)$ is the number of individuals in the neighborhood U_a of a selecting color λ_j at time step t. The absorbing states of this process are the monochromatic states; there are, of course, m absorbing states. It follows from the general results for Markov chains that the process will reach an absorbing state with probability one.

We first consider the case where the neighborhood of an individual is the entire audience, $U_a = A$. The parameter α is the relative size of the neighborhood so in this case

$$\alpha = \frac{\text{card}(U_a)}{N} = 1 . \qquad (2)$$

Formally, this case may be treated as a Markov chain. [6] The average time to absorption, $E\tau$, can be expressed in terms of iterations of the transition matrix. For example, for $m = 2$ colors, the transition matrix is

$$P_{ij} = \binom{N}{j} (i/N)^j (1-i/N)^{N-j} . \tag{3}$$

There are two absorbing states, $i=0$ and N so $P_{0j} = \delta_{0j}$ and $P_{Nj} = \delta_{Nj}$. We define the matrix $\mathbf{V} = (\mathbf{I} + \mathbf{Q} + \mathbf{Q}^2 + ...) = (\mathbf{I} - \mathbf{Q})^{-1}$ where $\mathbf{Q} \equiv \| P_{ij} \|$, $i,j=1,...N-1$. Then

$$E\tau = \sum_{j=1}^{N-1} \mathbf{V}_{ij} \tag{4}$$

where $i = xN$ is the initial state. Unfortunately, this formal solution becomes impractical to implement for $m \geq 4$ colors if N is not small (i.e. audiences with more than ten members).

2.2 Two Color Problem in the Diffusion Approximation

Since we are interested in asymptotic behavior for large N we introduce the diffusion process as a limit of discrete Markov chains. In order to do so we consider the sequence of chains with $N \to \infty$ and $\Delta t \to 0$ with the appropriate relation between N and Δt. In this sub-section we treat the two color ($m = 2$) case. The process starts from the state $\{x, 1-x\}$ where x is the fraction of voters initially selecting color λ_1. Consider the one dimensional diffusion process $\xi(t)$ ($\xi(0)=x$), where $\xi(t)$ is the fraction of voters selecting color λ_1 at time t. Since we take $m=2$, $1-\xi(t)$ is the fraction of voters selecting color λ_2 at time t.

This process should be homogeneous in time, therefore we can consider this process starting from any moment in time (say $t=0$). Let $T(x)$ be the average time to absorption, defined as

$$T(x) = E\, \tau(x) \tag{5a}$$

where

$$\tau(x) \equiv \inf\{t \geq 0 \mid \xi(0) = x, \xi(t)=0 \text{ or } \xi(t)=1 \} . \tag{5b}$$

The random variable $\tau(x)$ is bounded with probability one. We may also write the average time to absorption in the following form by averaging with respect to the first timestep,

$$T(x) = E \, \tau(x) = \Delta t + E\tau(\xi(\Delta t)) + o(\Delta t) \tag{6}$$

or

$$-1 = \frac{E\tau(\xi(\Delta t)) - T(x)}{\Delta t} + \frac{o(\Delta t)}{\Delta t} . \tag{7}$$

The limit $\Delta t \to 0$ in (7) is defined by the infinitesimal operator [7],

$$A = a(x)\frac{d}{dx} + \frac{1}{2}b(x)\frac{d^2}{dx^2} \tag{8}$$

of the process $\xi(t)$ such that

$$-1 = AT(x) = a(x)\frac{d}{dx}T(x) + b(x)\frac{d^2}{dx^2}T(x) \tag{9}$$

where

$$a(x) = \lim_{\Delta t \to 0} \frac{E(\xi(\Delta t) - x)}{\Delta t} \tag{10a}$$

$$b(x) = \lim_{\Delta t \to 0} \frac{E(\xi(\Delta t) - x)^2}{\Delta t} . \tag{10b}$$

By the description of the process $\xi(t)$, $N\xi(\Delta t)$ is binomial distributed with parameter x. Hence, $E\xi(\Delta t) = x$ and

$$E(\xi(\Delta t) - x)^2 = \text{Var } \xi(\Delta t) = \frac{x\,(1-x)}{N} . \tag{11}$$

Consequently,

$$a(x) = 0 , \quad b(x) = \frac{x\,(1-x)}{\Delta t\, N} . \tag{12}$$

If we choose $\Delta t = 1/N$ for $N \to \infty$ we have the following equation for $T(x)$,

$$-1 = \frac{1}{2}x\,(1-x)\frac{d^2 T}{dx^2} \tag{13}$$

with the boundary conditions $T(0) = T(1) = 0$. The solution of (13) is,

$$T(x) = -2\,[\,x \ln x + (1-x)\ln(1-x)\,] . \tag{14}$$

The asymptotic expression for the time to absorption in terms of Δt is,

$$\frac{T(x)}{\Delta t} = -2N\ [\ x \ln x\ +\ (1-x)\ \ln(1-x)\ ]+o(N)\ .\tag{15}$$

Interestingly, the above expression contains the information entropy of the random variable x.

2.3 Multi-color Problem in the Diffusion Approximation

The approach described in the previous sub-section cannot be directly applied to the multi-color case. In this case the absorption states do not coincide with the boundary of the corresponding problem. The average time till absorption can be defined as $E\tau(x_1,...,x_{m-1})$ where

$$\tau(x_1,...,x_{m-1}) = \inf\{t \geq 0 \mid \xi_1(0)=x_1,\ ...,\ \xi_{m-1}(0)=x_{m-1},$$
$$\xi_1(t)=1\ or\ ...\ or\ \xi_{m-1}(t)=1\ or\ \xi_1(t)+...+\xi_{m-1}(t)=0\ \}\ .\tag{16}$$

For the corresponding m-1 dimensional diffusion process, $\eta(t) = (\xi_1(t),...,\xi_{m-1}(t))$, $\eta(0) = (x_1,...,x_{m-1})$. As before, the drift term, a, vanishes and the *infinitesimal* operator A contains only the diffusion term, b, in the form,

$$A_{m-1} = \frac{1}{2}\sum_{ij=1}^{m-1} b_{ij}\frac{\partial^2}{\partial x_i\,\partial x_j}\ .\tag{17}$$

Here

$$b_{ij} = \lim_{N\to\infty}\frac{1}{\Delta t}\ E(\xi_i(\Delta t)-x_i)(\xi_j(\Delta t)-x_j)\tag{18}$$

with $\Delta t=1/N$. It can be easily derived that

$$b_{ij} = \begin{cases} x_i\,(1-x_i) & i=j \\ -x_i\,x_j & i\neq j \end{cases}\tag{19}$$

because the vector $N\eta(\Delta t)$ is multinomial distributed with parameters $x_1,...,x_{m-1}$ and therefore

$$E(\xi_i(\Delta t)-x_i)\ =\ 0\tag{20a}$$

$$E(\xi_i(\Delta t)-x_i)^2\ =\ Var\ \xi_i(\Delta t)\ =\ \frac{x_i\,(1-x_i)}{N}\tag{20b}$$

$$E(\xi_i(\Delta t) - x_i)(\xi_j(\Delta t) - x_j) = \text{Cov } \xi_i(\Delta t)\xi_j(\Delta t) = \frac{-x_i x_j}{N}. \tag{20c}$$

The boundary value problem for the average first passage time is similar to (13), for the m-1 dimensional case; it may be written in the form

$$-1 = A_{m-1} T^*(x_1,...,x_{m-1}) \quad ; \quad (x_1,...,x_{m-1}) \in D_{m-1} \tag{21}$$

and $T^*\big|_{\partial D_{m-1}} = 0$. The trajectories lie within the m-1 dimensional simplex

$$D_{m-1} = \{ (x_1,...,x_{m-1}) \in \mathfrak{R}^{m-1} \mid x_i \geq 0, \sum_{i=1}^{m-1} x_i \leq 1 \}. \tag{22}$$

However, we see that the solution $T^*(x_1,...,x_{m-1})$ of this boundary value problem differs from $T(x_1,...,x_{m-1})$ because the two do not satisfy the same boundary conditions except when $m=2$. For example, for $m=3$, if we take an initial state on the boundary, such as $(x_1,x_2=0)$, then $T^*(x_1,x_2=0) = 0$. On the other hand, $T(x_1,x_2=0) \neq 0$ since $(x_1,x_2=0)$ is <u>not</u> an absorbing state except when $x_1 = 0$ or 1.

To use this approach we might consider the following sequence of processes: the $(i+1)^{\text{th}}$ process is defined in the domain which is the boundary of the domain of the ith process. The starting point for a process is the point of the first passage of the boundary for the previous process (see Fig. 1).

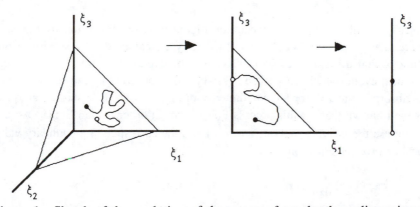

Figure 1: Sketch of the evolution of the process from the three dimensional phase space to the absorbing state. The solid circles are the initial states and the open circles are the final states in each step of the process.

The initial point for the 3-dimensional process $\eta_3(t) = (\xi_1(t), \xi_2(t), \xi_3(t))$ is $\eta_3(0)=M_0$ (dark circle in the left figure). The trajectory $\eta_3(t)$ in D_3 reaches the boundary D_2 at the point M_1 (open circle in the left figure; dark circle in the middle figure). From this point we define a 2-dimensional process $\eta_2(t) = (\xi_1(t), \xi_3(t))$, $\eta_2(0)=M_1$. Similarly, this trajectory reaches the line D_1 and the trajectory on that line eventually reaches the absorbing state $(0,0,0)$ (Note: the points $(0,0,1)$, $(0,1,0)$ and $(1,0,0)$ are also absorbing states). According to this approach we could obtain the average time it takes the process to reach one of the surfaces of the simplex starting from M_0 by solving the corresponding boundary value problem (21) with $m=4$. The next step is problematic because for the $m=3$ problem the initial condition $\eta_2(0)=M_1$ is a random variable with a distribution defined by the corresponding Fokker-Planck equation.

We may avoid these difficulties by employing the following trick; consider the sequence of events,

$$C_1 = \{\xi_1(\tau)=1\}, \ldots ,C_{m-1} = \{\xi_{m-1}(\tau)=1\},$$

$$C_m = \{\xi_1(\tau)+\ldots+\xi_{m-1}(\tau)=0\} \qquad (23)$$

and $C = \overset{m}{\underset{i=1}{U}} C_i$. Obviously, the events C_i are mutually exclusive. The expected time to absorption may be written as

$$T(x_1,\ldots,x_{m-1}) \;=\; E\tau \;=\; \sum_{i=1}^{m} E(\tau|C_i)\, P(C_i) \;. \qquad (24)$$

$E(\tau|C_i)$ is the conditional expectation of the time to absorption by the $x_i=1$ state. The values of $E(\tau|C_i)$ and $P(C_i)$ may be calculated by solving a one dimensional diffusion problem. Consider the event C_i together with its complement event (which has probability $1-P(C_i)$). The event C_i can be interpreted as absorption at $x=1$ for the one-dimensional process on $[0,1]$ starting at x_i . It is well known that for this process $P(C_i) = x_i$. In order to find $E(\tau|C_i)$ we can again use the infinitesimal operator (17) and the corresponding boundary value problem for the equation

$$\frac{1}{2} x_i\, (1-x_i)\, \frac{\mathrm{d}^2}{\mathrm{d}x_i{}^2}\, g(x_i) \;=\; -x_i \qquad 0 < x_i < 1 \qquad (25)$$

where $g(x_i) = x_i\, E(\tau|C_i)$ and the boundary conditions are $g(0) = g(1) = 0$. The solution of this boundary value problem is

$$g(x_i) = -\frac{2(1-x_i)\ln(1-x_i)}{x_i} \qquad i=1,\ldots,m-1 \;. \qquad (26)$$

For $i=m$ the solution may be written in the same form if we take $x_m = 1 - x_1-\ldots-x_{m-1}$. Combining (24) and (26) we have

$$T(x_1,\ldots,x_{m-1}) = -2\sum_{i=1}^{m}(1-x_i)\ln(1-x_i)\;. \qquad (27)$$

In terms of the length of the characteristic time interval Δt, the time until "unison" can be asymptotically expressed in the form

$$\frac{T(x_1,\ldots,x_{m-1})}{\Delta t} = -2N\sum_{i=1}^{m}(1-x_i)\ln(1-x_i) + o(N) \qquad (28)$$

Finally, consider the case where the each member of the population initially selects a color at random with equal probability. The initial distribution of colors is then multinomial. The mean value of T with respect to the initial distribution will be,

$$\frac{\langle T\rangle}{\Delta t} = -2N\sum_{i=1}^{m}\langle(1-x_i)\ln(1-x_i)\rangle + o(N) \qquad (29)$$

where

$$\langle(1-x_i)\ln(1-x_i)\rangle = \sum_{k=0}^{N}\binom{N}{k}\frac{1}{m^k}(1-\frac{1}{m})^{N-k}(1-\frac{k}{N})\ln(1-\frac{k}{N})\;. \qquad (30)$$

The function $x\ln x$ is continuous in $[0,1]$ therefore using Bernstein's theorem [8],

$$\lim_{N\to\infty}\sum_{k=0}^{N}\binom{N}{k}p^k(1-p)^{N-k}f(k/N) = f(p)\;. \qquad (31)$$

Using this result in (30) we have

$$\frac{\langle T\rangle}{\Delta t} \approx -2N\,(m-1)\ln(1-\frac{1}{m}) \qquad (32)$$

in the limit of large N and

$$\frac{\langle T \rangle}{\Delta t} \approx 2N \left(1 - \frac{1}{m}\right) \tag{33}$$

in the limit that both N and m are large.

3. NUMERICAL RESULTS

3.1 Monte Carlo Simulation

A Monte Carlo program was used to study the behavior of this model in a variety of scenarios. For each trial, the program assigns an initial color to every site. Since each color is equally probable the initial probability distribution is multinomial. From a site's neighborhood a member is chosen at random and the color of the site is reset to that of the selected neighbor. Note that a site is a member of it's own neighborhood. The algorithm updates the state of all sites in parallel so it is in the class of stochastic cellular automata. When the system reaches the absorbing state of unison we record the time and start a fresh trial.

The results presented in the following sub-sections are ordered according to the dimensionality of the system (see Fig. 2).

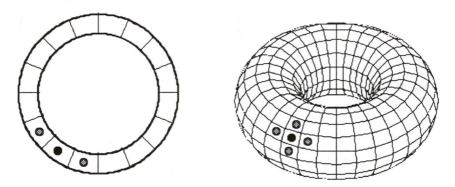

Figure 2: Schematic illustration of one and two dimensional audiences with periodic boundary conditions. For the site indicated with a black circle, the nearest neighbors are indicated by shaded circles. Note that a site's neighborhood includes the site itself so in the two figures shown above $n = 3$ and 5, respectively.

3.2 Infinite Dimensional System ($\alpha = 1$)

We first consider the case where the neighborhood of each site contains all other sites (or in the analogy of a clapping audience, everyone hears everyone else). This case may be considered to be the infinite dimensional system. The average time to absorption is given by (29) or, in the limit of large N, by (32). Note that the time increases linearly with N and the results from the Monte Carlo program are in quantitative agreement with the theory, as shown in Fig. 3 and 4.

From (33), we see that as the number of colors, m, increases the mean absorption time increases but only as $(1 - 1/m)$. In Fig. 5 we plot the average time to absorption as a function of m and indeed see that it depends weakly on the initial number of colors.

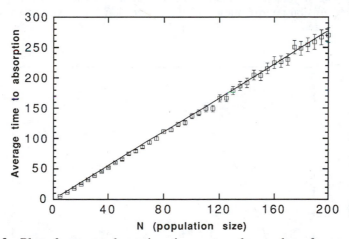

Figure 3: Plot of average absorption time versus the number of persons in the audience for $m=2$ colors. The data points are the results from the Monte Carlo simulation. Each point represents an average over $2 \cdot 10^5/N$ trials (e.g. for $N=200$ persons there were 1000 trials). The solid line is given by (29), the diffusion approximation to the absorption time. See the next figure for results from multi-color systems.

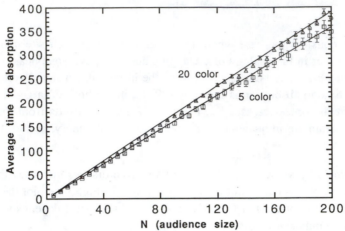

Figure 4: Plot of average absorption time versus the number of persons in the audience for $m=5$ and 20 colors. The data points are the results from the Monte Carlo simulation. Each point represents an average over $2 \cdot 10^5/N$ trials (e.g. for $N=200$ persons there were 1000 trials). The solid line is given by (29), the diffusion approximation to the absorption time. Compare with the previous figure.

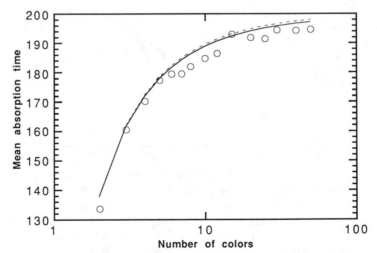

Figure 5: Plot of mean absorption time versus initial number of colors, m, for an audience size of $N = 100$. The data points are the results from the Monte Carlo simulation. Each point represents an average over 10^4 trials. The solid line is (29) and the dashed line is it's approximation, (32).

3.3 One-dimensional System

For a one-dimensional system, the average time to absorption depends on the number of members in the neighborhood, n (see Fig. 2). When the neighborhood size is comparable to the audience size we expect to recover the infinite dimensional results described above. When the neighborhood consists of only the nearest neighbor sites (left, right and center, see Fig. 2) then the average time to absorption goes roughly as N^2 (see Fig. 6).

The evolution of a typical trial is illustrated in Fig. 7. Note that after an initial transient the system naturally divides itself into two well defined regions of different color. This occurs even when the initial distribution is multi-color since, for large N, the system quickly evolves to a two color state. An absorbing state is reached when the two boundaries of one of the regions meet.

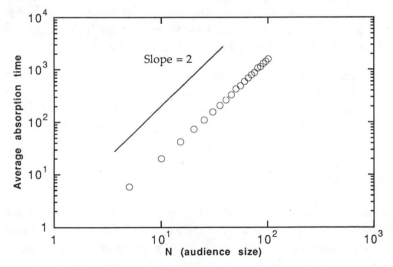

Figure 6: Plot of mean absorption time, versus audience size, N, for the case of $m = 2$ colors. The system is one dimensional with periodic boundary conditions. The neighborhood consists of three nearest neighbor sites (left, right and center). The data points are the results from the Monte Carlo simulation. Each point represents an average over 2000 trials The solid line has a slope of two; a least squares fit of the data gives a slope of 1.89.

time

Figure 7: Illustration of a typical evolution to unison in a one-dimensional system with periodic boundary conditions. The neighborhood consists of three nearest neighbor sites (left,right and center). The audience size is $N=80$ and the time to absorption is 616 timesteps. The system starts with a random mix of the two colors (far left column) and reaches an absorbing state when all sites are white (far right column). Note that for most of the evolution the system is divided into a two regions.

In this picture, we may map the problem into a first passage time problem for a random walk. Consider a pair of random walkers initially placed at random on a one-dimensional grid of length N with periodic boundary conditions. The average time for the walkers to meet goes as N^2, which agrees qualitatively with the results shown in Fig. 6.

The parameter $\alpha = n/N$ is the relative size of the neighborhood. The average time to absorption varies with α but, as illustrated in Fig. 8, the dependance is very weak for $\alpha > 0.1$. On the other hand, for $\alpha < 0.1$, the mean absorption time appears to follow a power law dependance.

3.4 Two-dimensional System

At present we have only very preliminary results for two-dimensional systems. Figure 9 illustrates a large audience at two stages of the evolution. Notice that while large, monochromatic regions develop the system retains small scale structures. In fact, these small scale structures persist even as the evolution approaches the absorbing state. As shown in Fig. 10, the mean absorption time is approximately linear with N when the neighborhood is von Neumann (see Fig. 2).

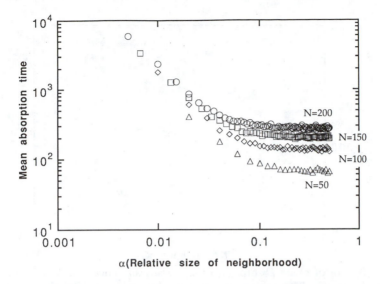

Figure 8: Plot of mean absorption time, versus neighborhood size, α, for the case of $m = 2$ colors and various audience sizes. The system is one dimensional with periodic boundary conditions. Each point represents an average over 2500 trials in the Monte Carlo simulation.

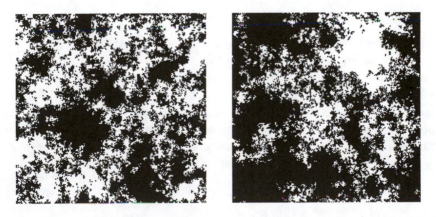

Figure 9: Snapshots of the system state for a two dimensional audience of 40,000 members (periodic boundary conditions). The picture on the left is after 10^3 iterations; the picture on the right is after 10^4 iterations. Estimated time to absorption is about 10^5 iterations.

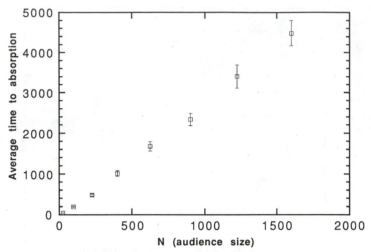

Figure 10: Plot of mean absorption time, versus audience size, N, for the case of $m = 2$ colors. The system is two dimensional with periodic boundary conditions. The neighborhood consists of only the five nearest neighbor sites (von Neumann neighborhood). The data points are the results from the Monte Carlo simulation. Each point represents an average over 100 trials.

4. FUTURE WORK

At present we are further investigating the influence of neighborhood size and geometry on the absorption time. For one-dimensional systems we are attempting to determine whether there is a "critical" value of $\alpha \approx 0.1$ at which the relation between absorption time and audience size structurally changes (see Fig. 8). For both one- and two-dimensional systems we would like to obtain analytic results similar to those discussed in section 2 for the infinite dimensional system. Finally, to investigate the effects of neighborhood geometry we are numerically investigating the model on d-dimensional lattices (e.g. percolation lattice, Sierpinski carpet).

The authors would like to thank Jim Garahan, Hirsh Lev, Ron Tomas and Susan Hansen for helpful discussions.

References

1.   R.A. Fisher, *The Genetical Theory of Natural Selection*, Oxford Univ. Press, New York (1930); S.Wright, Proc. Natl. Acad. Sci. U.S., **31**, 382 (1945); M.Kimura, Ann. Math. Statist. **28**, 882 (1957).

2.   M. Feldman, in *Lectures in the Sciences of Complexity*, ed. D.L. Stein (Addison-Wesley, Redwood City, 1989).

3.   M. Malek Mansour and A. De Palma, Physica **128A**, 377 (1984).

4.   M. Kimura and G. Weiss, Genetics **49**, 561 (1964); G. Vichniac, in *Disordered Systems and Biological Organization*, eds. Bienenstock et. al. (Springer, Berlin, 1986) pg. 1.

5.   R. Balescu, *Equilibrium and Nonequilibrium Statistical Mechanics* (Wiley, New York, 1975); S.K. Ma, *Statistical Mechanics* (World Scientific, Teaneck, 1985).

6.   J. Kemeny and J. Snell, *Finite Markov Chains* (Van Nostrand, Princeton, 1960); A.T. Bharuch-Reid, *Elements of the Theory of Markov Processes and their Applications* (McGraw-Hill, New York, 1960).

7.   C.W. Gardiner, *Handbook of Stochastic Methods* (Springer, Berlin, 1990).

8.   W. Feller, *An Introduction to Probability Theory and its Applications*, Vol. II (Wiley, New York, 1966).

Nonhomogeneous Response of Reaction-Diffusion Systems to Local Perturbations

B. Cayco, M. Feldman and V. Kanevsky

We investigated the behavior of the Brusselator, a model of a biochemical reaction. We will show that the system's response to local perturbations depends upon the points where the perturbations have been applied. We have also discovered the universality of the approach to a stable solution along the trajectory of this system.

1. INTRODUCTION

The existence of stable nonhomogeneous states in the mathematical model of open chemical reactions has been known since Alan Turing's pioneering work [1]. Later, this phenomenon has been observed in models motivated by problems arising in various fields such as biology, sociology, economics, etc., and eventually, the chemical reaction with Turing structures has been discovered [2] (see also [3] and [4]). One can expect that considerations of relatively simple models of biochemical systems can be used to achieve a better understanding of the variety of phenomena present in living organisms. Since all the objects considered in synergetics belong to the class of open systems, the study of reactions under the influence of external action (the presence of external actions is characteristic for open systems) is entirely natural.

We are studying the sensitivity of different points of the nonhomogenous stable states with respect to external actions. Interest in this problem was triggered by the association with the phenomenon of acupuncture points on a human body. Any living system can be considered as a hierarchy of

B. Cayco and V. Kanevsky: Department of Mathematics and Computer Science, San Jose State University, San Jose, CA 95192

M. Feldman: Decision Focus Incorporated, 4989 El Camino Real, Los Altos, CA 94022

biochemical reactions. The existence of extremely sensitive points was discovered on many animals. A natural way to model this phenomenon is a reaction-diffusion system (RDS) and perturburbations to their steady states. The following question immediately arises while trying to describe this effect: what is the lowest level of organization that manifests this peculiarity? As we shall see, a model of simple chemical reactions already has this property.

There are two questions we would like to answer. First, we will investigate the dynamics of the development of perturbations to a reaction-diffusion system and the time evolution of perturbation under impact depending on type and character of the structure, and also on perturbation's location, magnitude, etc. Second, we will investigate the transition between spatial structures under local and global external action; switching between the structures by finite duration perturbations of the dynamical variables.

Our results are based on a computer simulation of a reaction-diffusion system (RDS). An inherent concern pertains to the validity of our conclusions which where derived from the numerical simulation of a particular RDS. This question can be posed as follows: how conclusive are our observations and analysis regarding general properties of an RDS? The resolution of this question (on the physical level or rigor) is based on the natural scientific ideas concerning the universality of certain fundamental properties of an RDS which can be found in numerous known RDS models. We enumerate some of these properties.

(i) The existence of spatial nonhomogeneous solutions of an RDS and solutions of the type of time-dependent oscillations.
(ii) The existence of bifurcations as a result of variations in control parameters.
(iii) The diversity of stable attractors corresponding to the same boundary conditions and possibility for them to switch from one to the other under external impact.

2. DESCRIPTION OF THE MODEL

As our prototype for an RDS, we have chosen the Brusselator, a mathematical model for a set of chemical reactions (1) based on the "Law of Mass Action".

$$
\begin{aligned}
A &\to X \\
B + X &\to Y + D \\
2X + Y &\to 3X \\
X &\to E
\end{aligned}
\tag{1}
$$

where A, B, D, E, X and Y are symbols of reacting substances. Non-equilibrium conditions are created by immediate elimination of the products D and E of the reaction. Concentrations of A and B are kept constant while the concentrations of the substances X and Y are allowed to evolve. The RDS described by (1) can be expressed as the initial-boundary value problem for the following reaction-diffussion equations:

$$\frac{\partial X}{\partial t} = A - (B + 1)X + X^2 Y + D_X \nabla^2 X$$

$$\frac{\partial Y}{\partial t} = BX - X^2 Y + D_y \nabla^2 Y$$

$$(2)$$

where $X = X(t, r)$, $Y = Y(t, r)$ are the concentrations of substances X and Y respectively, t is time, r is the space variable, A, B are constants, and D_X, D_y are the diffusion coefficients for the substances X and Y respectively.

We are considering the one-dimensional reactor with $0 \leq r \leq 1$ and the following boundary non-flow conditions:

$$\frac{\partial X}{\partial r}\bigg|_{r = 0, 1} = \frac{\partial Y}{\partial r}\bigg|_{r = 0, 1} = 0$$

$$(3)$$

and the initial conditions

$$X(0, r) = X_0(r), \quad Y(0, r) = Y_0(r).$$

Linear analysis for the loss of stability of the steady state solution:

$$X(t,r) \equiv A, \quad Y(t,r) \equiv B/A.$$

can be done by using standard techniques of perturbation theory.

When the parameter B exceeds a critical value, B_T, the steady state solution looses its stability and stable nonhomogeneous solutions may appear. This is known as the Turing bifurcation. It can be shown that the critical value is given by (see [5] and [6]):

$$B_T = (1 + \eta A)^2, \text{ where } \eta = (\frac{D_x}{D_y})^{\frac{1}{2}}.$$

3. FORMULATION OF THE PROBLEMS AND THE METHODOLOGY OF NUMERICAL EXPERIMENTS

What follows is the study of the evolution of an RDS under external impact, by direct action onto dynamical variables X and Y. For numerical integration of the system (2) and (3), the following implicit scheme

$$\frac{x_m^{n+1} - x_m^n}{\Delta t} = f(x_m^n, y_m^n) + D_x \frac{x_{m-1}^{n+1} - 2x_m^{n+1} + x_{m+1}^{n+1}}{h^2}$$

$$\frac{y_m^{n+1} - y_m^n}{\Delta t} = g(x_m^n, y_m^n) + D_y \frac{y_{m-1}^{n+1} - 2y_m^{n+1} + y_{m+1}^{n+1}}{h^2} \tag{4}$$

was implemented. Here $f(x,y) = A - (B+1)x + x^2 y$ and $g(x,y) = Bx - x^2 y$, Δt is the time step, h is the space step, X_m^n is the approximation of $X(n\Delta t, mh)$. The boundary conditions are:

$$x_0^n = x_1^n, \quad y_0^n = y_1^n; \quad x_{M-1}^n = x_M^n, \quad y_{M-1}^n = y_M^n \quad (n = 0,1,...) \tag{4'}$$

and the initial conditions are:

$$x_m^0 = x_m, \quad y_m^0 = y_m \quad (m = 0,...,M) \tag{4''}$$

where x_m and y_m are arbitrary. Let the vector, $\mathbf{x}^n = \{x_m^n\}_{m=0}^M$, represent the solution of the system at time $t = n\Delta t$.

In some of the numerical experiments, a stable solution, $\mathbf{x}^0$, was perturbed at certain points and the system was allowed to evolve back to the stable solution $\mathbf{x}^0$. At each time step, two deviations, d_C and d_{L_2}, are calculated. These two quantities measure the distance of the current state, $\mathbf{x}^n$, from the original solution.

The quantities d_C and d_{L_2} are defined as follows:

$$d_c = \max \{|x_m^n - x_m^0|, \, m = 1,2,..., M\}$$

$$d_{L_2} = \sqrt{\frac{1}{n} \sum_m (x_m^n - x_m^0)^2} \tag{5}$$

The convergence of the numerical procedure (4, 4') with certain initial conditions is controlled by the magnitude of the deviations d_C and d_{L_2} between the vectors $\mathbf{x}^n$ and a certain fixed vector which, in particular, might be a steady state solution. In some cases, the vector whose evolution has been studied was selected as such a fixed vector.

The calculations were stopped when K significant digits of d_C and d_{L_2} remained unchanged. The value of K, which depends on Δt and h, was heuristically chosen during the computational experiments. (We used K = 6.) The implicit difference scheme described above is absolutely stable for the linear heat equation, i.e., it is stable for any choice of Δt and h [3]. In our case, the scheme is no longer absolutely stable because of the nonlinearity introduced by the functions f and g. For fixed h, a smaller value of Δt has to be chosen when the values of the control parameters (A, B, D_x, D_y) are close to their bifurcation values. The value of h depends upon the shape of the stable solution; we chose h sufficiently small in order to capture the extreme points. Roughly speaking, we alloted 8-10 grid points per peak.

4. DYNAMICS OF PERTURBATIONS OF DISSIPATIVE STRUCTURES

It is known that a whole family of dissipative structures (DS) can satisfy the system (2) and (3) with the same values of the control parameters [6]. In Fig. 1., we show two structures that were used for numerical experiments.

Now , let us consider the peculiarities of the evolution of local perturbations. Perturbations of a given magnitude were applied locally and the RDS was allowed to evolve. The quantities , d_C and d_{L2} , were calculated at each time step and the results were plotted. This proceedure was repeated for perturbations at the extreme points, inflection points and other points of the

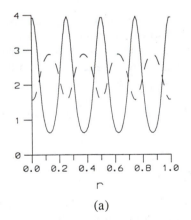

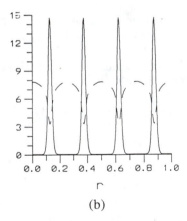

(a)   (b)

Fig. 1. The stable solutions of (2) and (3).   (a) B = 5.0, h = 0.0125, Δt = 0.01. (b) for B = 40.0, h= 0.0025, Δt = 0.0005. The solid line (——) represents X and the dashed line (- - -) represents Y. For both figures, A = 2.0, D_x = 0.0016, and D_y = 0.008.

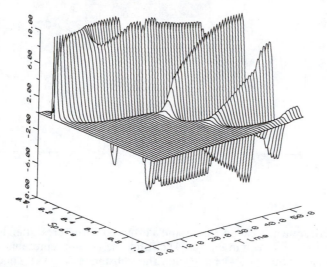

Fig. 2. Time evolution of the normalized difference between the perturbed and initial structures for A = 1.0, B = 1.5, D_X = 0.01, and D_y = 0.0001.

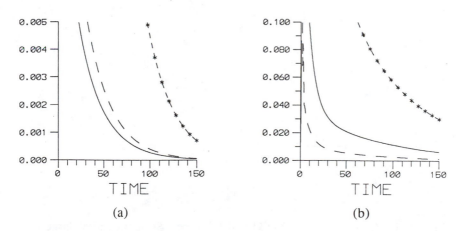

<table>
<tr><td>(a)</td><td>(b)</td></tr>
</table>

Fig. 3. Relaxation Rates (d_C) for Different Initial Perturbations. The initial states for Figs. 3a and 3b are those from Figs. 1a and 1b, respectively. The solid line (———) represents the DS which was perturbed at minimum points, (- - -) represents the DS which was perturbed at maximum points, and (-*-*-) represents the DS which was perturbed at inflection points. The magnitude of the initial perturbations are the same for all cases.

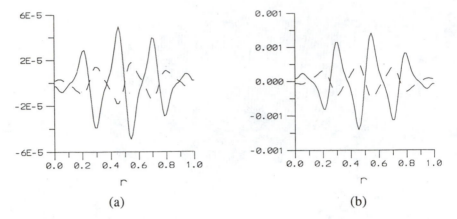

(a)

(b)

Fig. 4. Difference between Initial and Perturbed Structures after large time
The initial state is shown in Fig 1a. The solid line (———) represents substance
X and dashed line (- - -) represents the substance Y. (a) This DS was
perturbed at minimum points. (b) This DS was perturbed at an inflection
point.

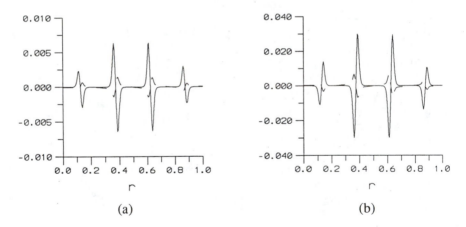

(a)

(b)

Fig. 5. Difference between Initial and Perturbed Structures after large time
The initial state is shown in Fig 1b. The solid line (———) represents substance
X and dashed line (- - -) represents the substance Y. (a) This DS was
perturbed at minimum points. (b) This DS was perturbed at an inflection
point.

given structure. Our computer experiments show that the time of relaxation depends on the point where the perturbation has been applied. Perturbations applied to the point of inflection of a quasi-harmonic DS die out slower then those applied to any other point (See Fig. 3). We also performed similar numerical experiments on structures other than those in Fig. 1 and arrived at the same results.

Our calculations indicate that the points of inflection play a very significant role in the time evolution of the perturbed RDS. Perturbations at these points are the slowest to die out. Thus, one can see that the inflection points have the highest sensitivity with respect to perturbations.

Another observation is that the local maxima of absolute values of deviations of the solutions of (2) and (3) from the corresponding limiting DS's are attained at the points of inflection of this DS and that the zeros of the deviations coincide with the local extrema of the original DS. In Figs. 4 and 5, we show the examples of deviations from the structures shown in Fig. 1. Numerical experiments with many different perturbations of several structures showed the same result. This type of asymptotic behavior is independent of the character of perturbations: it can be local or integral, it can be switching or not switching, etc. This problem could be of interest for analytical treatment and this has not yet been done.

5. SWITCHING OF REGIMES OF FUNCTIONS OF RDS BY EXTERNAL ACTION

For any given values of the parameters, A, B, D_x, and D_y, there is a family of the stable stationary solutions of the RDS. Each member of this family is an attractor of dynamical system and has its own domain of attraction. We, therefore, ask the following questions: Is it possible for a stable solution to evolve to another stable solution under external action which belongs to some class of perturbations? If so, what are the appropriate forces that induce the system to switch to a different state when only the original state is known? We present a partial answer to this question. We considered the evolution of the DS under perturbations of the following kind.

Denote by X(r) and Y(r) the nonhomogeneous stationary solutions of (2) and (3) and let

$$\delta X_i(t,r) = \begin{cases} -X(r) & |r-r_i| < \varepsilon, 0 \le t \le T \\ 0 & \text{otherwise} \end{cases}$$

where r_i is a maximum of X(r), T is some non-negative constant. T is another characteristic of this kind of perturbation. The switching diagram for

292

the contrast structures for A = 1, B = 1.5, D_x = .01, D_y = .0001 is shown in Fig. 6.

The cycle generated by this switching mechanism might not include all the members of the family of DS. This indicates the limitation of this kind of perturbations with respect to the switching phenomenon. The switching phenomenon under local impact on DS has been discovered in the framework of each family of DS's.

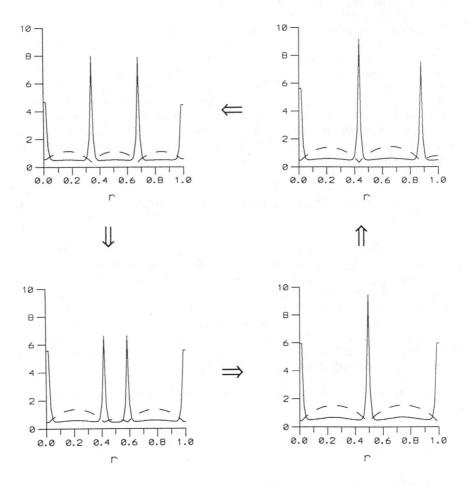

Fig. 6. Switching between contrast structures under space-time perturbations for A = 1.0, B = 1.5, D_x = 0.001, and D_y = 0.01. The solid line (——) represents X and the dashed line (- - -) represents Y.

6. CONCLUSIONS

First of all, let us emphasize the fact that a simple model of an RDS (Brusselator) manifests a manifold of nontrivial settled time regimes of behavior and stationary spatial structures (DS). The families of these regimes can differ by the number of maximum points; they can be quasi-harmonic or contrast. Near bifurcation points intricately periodical intermediate regimes usually appear. They exist for a comparatively long period of time and then transfer to settled regimes of oscillations or spatial structures. The measured external action can change any settled regime of RDS. Some characteristic properties of the system's response to those types of impact which were considered are:

(i) The possibility to switch the regime of RDS by applying perturbations from some restricted class of functions.

(ii) The dependence of the action's result on its location. The existence of spatially peculiar points of the system where the action is especially effective.

(iii) The asymptotic behavior of the system in the neighborhood of stationary state: The maximum of absolute values of deviations are attained at the inflection points of a final stationary structure.

The study of the reaction of an RDS to external action was undertaken as a step towards a solution of general problem of "control" of a nonlinear system. The term "control" can be interpreted as goal-driven change of the state of RDS, i.e., a sequence of external actions which would lead the system to a desired final state. The solution of this apparently extremely complicated problem ought to start with more precise formulation of the problem. It is necessary to notice that if one considers the class of all possible external actions (assuming that all the attractors of the system are known!), the solution of the problem of control becomes trivial. Therefore, a reasonable formulation of the problem ought to contain natural restrictions on the "dimension" of the set of all possible "control" influences. In the above-mentioned case, this dimension is infinite. For example, the number of independent parameters which define all class of possible actions can be considered as a dimension. So, if one wants to limit the class of external actions by perturbation only to one parameter of RDE at one point, we obtain a one-dimensional space with the coordinate being equal to the magnitude of the action. The same scheme for two points leads to the two-dimensional space. Thus, all external actions considered in this paper can be included in the corresponding finite dimensional space.

In connection with these considerations, the general problem of control can be reformulated more accurately as a search for such a class of control actions which would minimize the number of parameters needed for their description and still allowing the switching to any desirable state. We can consider further restrictions of this control problem by limiting some characteristics of actions which can be expressed in terms of functionals of trajectories of the system.

These functionals usually correspond to some physical characteristics of the system: energy, magnitude, duration, etc.

We would like to thank Professor Vladimir Naroditsky for his illuminating discussions and valuable suggestions. We are grateful to the referee for enlightening us about the recent discovery of Turing type patterns in a chemical reaction.

REFERENCES

1. A. M. Turing,"The Chemical Basis of the Morphogenesis", Proc. Roy. Soc.B **237**, 37-71, (1952).

2. V. Castets, E. Dulos, J. Boissonade, and P. de Kepper, Phys. Rev. Lett. **64**, 2953 (1990).

3. I. Lengyel and I. R. Epstein, "Modeling of Turing Structures in the Chlorite-Iodide-Malonic Acid-Starch Reaction System", Science **251**, 650-652, (1991).

4. A. Arneodo, J. Elezgaray, J. Pearson and T. Russo, "Instabilities of Front Patterns in Reaction-Diffusion Systems", Physica D **49**, 141-160, (1991).

5. P. Glansdorff and E. Prigogine, *Thermodynamics Theory of Structure, Stability and Fluctuations* (Wiley, New York, 1971).

6. G. Nicolis and E. Prigogine, *Self-Organization in Non-equilibrium Systems* (Wiley, New York, 1977).

7. J. M. Ortega and W. G. Poole Jr, *An Introduction to Numerical Methods for Differential Equations* (Pittman, Marshfield, 1981).

8. Yu. M. Romanovskii, N. V. Stepanova, and D. S. Chernavskii, *Mathematical Biophysics* (Nauka, Moskow, 1984).

Nonequilibrium Transient Near a Noise-Induced Instability

H.K. Leung

Nonlinear system of molecular self-replication is modeled with stochastic considerations. External noise is introduced through the fluctuating replication rate parameter. Treatment with white noise formulation basing on Stratonovich interpretation, together with approximation schemes enable us to derive the steady state as well as transient characteristics of the system. Transient processes near a noise-induced instability are found to exhibit *critical slowing-down* and attraction-basin shrinkage. The role played by the nonlinearity of system is discussed. *Nonlinear relaxation times* defined in terms of the first two moments are computed numerically, so that comparison can be made between them and the actual relaxation time.

1. INTRODUCTION

In modeling a complex system by using a set of phenomenological equations, we are in fact using a very small number of variables and parameters to represent an enormous number of degrees of freedom. This bears a consequence that all the variables and parameters describing the systems are averaged quantities, which are subject to intrinsic stochasticity. Furthermore, most of the nonequilibrium systems are open systems which interact strongly with environment. Stochastic treatment [1] is therefore essential in analyzing the effects on dynamic systems due to intrinsic fluctuations and extrinsic noises.

H.K. Leung : Institute for Nonlinear Science and Chemistry Department B-040, University of California at San Diego, La Jolla, CA 92093 U.S.A.

Permanent address: Physics Department, National Central University, Chung-li, Taiwan, 32054 ROC.

Transient process marks another difference between equilibrium and non-equilibrium systems. We shall study the stochastic transient processes in a simple nonlinear system, especially near its noise-induced instability. We are in fact modeling a stochastic transient preceding a nonequilibrium phase transition.

2. DETERMINISTIC MODEL AND INTRINSIC STOCHASTICITY

Our model system is derived from Eigen's model of macromolecular self-organization [2]. A species of molecules which are self-replicating with a rate W is constrained to have an ultimate concentration Ω. Dynamics of the system is described by a rate equation for the concentration x,

$$\frac{dx}{dt} = Wx - Wx^n/\Omega^{n-1}, \qquad n=2, 3. \tag{1}$$

In the above we consider only the case of $W > 0$ and $x \geq 0$. The equation with $n=2$ is similar to the logistic equation in Ecology; the one with $n=3$ is sometimes called the time-dependent Landau-Ginsburg equation in nonequilibrium Physics. They can also be related to Chemical and laser systems.

Deterministic transient can be solved from Eq.(1) analytically,

$$\xi(\tau) = \xi_0 \left[\xi_0^{n-1} + (1 - \xi_0^{n-1}) e^{-(n-1)\tau} \right]^{-1/(n-1)}, \tag{2}$$

where $\xi = x/\Omega$, and $\tau = Wt$. A system starts with $x_0 > 0$ will surely relax to a stable steady state (SS) with $x_s = \Omega$. Mathematically, the total *relaxation time* would be infinite. Practically, we may use an accuracy in measurement or in computation as a criterion of being at SS. For example, if a difference in $\xi_s - \xi(t) \approx 10^{-6}$ is considered as being at SS, a system with $\xi_0 \approx 0.1$ will assume a relaxation time $\tau_s \approx 15$ for $n=2$, and $\tau_s \approx 10$ for $n=3$.

Stochastically, the system is described by a probability function P(x,t). Due to nonlinearity of the system, P(x,t) is unlikely to be solved from the master equation [1]. Instead we look for the moments defined by,

$$\overline{x^k} = \int_0^\infty x^k P(x,t)\, dx . \tag{3}$$

It is well-known that the first two moments play predominant role in describing P(x,t). In a Guassian approximation, P(x,t) has its peak located at the mean $\overline{x}(t)$, and its width defined by the mean variance $\sigma(t) = \overline{x^2} - \overline{x}^2$. The infinite hierarchy of moment equations can be truncated by using the moment expansion scheme [3]. We expand all higher order moments by the first two,

$$\overline{x^k} \approx \overline{x}^k + \frac{k(k-1)}{2}\overline{x}^{k-2}\sigma .$$ (4)

We finally obtain a set of two equations in closed form,

$$\frac{d\overline{\xi}}{d\tau} = \overline{\xi}\left[1 - \overline{\xi}^{n-1} - \frac{n(n-1)}{2}\overline{\xi}^{n-3}\sigma\right]$$ (5)

$$\frac{d\sigma}{d\tau} = 2(1 - n\overline{\xi}^{n-1})\sigma - \Omega(\overline{\xi} - \frac{d\overline{\xi}}{d\tau}) .$$

Numerical results show that both moments will relax to a SS. We define this state as the *stochastic SS* since the approximated $P(x,t)$ is stationary in time. It is found that the relative fluctuation, $R = \sqrt{\sigma}/\overline{x}$, follows the law of large number. It is insignificant unless the system starts with x_0 too small as compared with Ω. This is a stochastic aspect of the size-effect: fluctuation catastrophe [2] might occur at the early stage of the replication process. We note that the extinction state with $x=0$ is an absorbing one, once it is reached there will be no recovery.

3. TRANSIENT NEAR NOISE-INDUCED INSTABILITY

Stochastic transient does show slowing-down phenomenon, sice that σ takes longer relaxation time than $\overline{x}$ does. It would be interesting to probe the transient behaviors under the influence of external noises, especially when the system is driven toward an instability by noises.

White-noise realization is basing on the assumption that noises are not correlated, since the correlation time is small as compared with the time scale of replication process. We consider a mixed order multiplicative noise. It arises as the replication rate is fluctuating randomly about its mean W with amplitude D,

$$W_t = W + D\zeta_t , \qquad <\zeta_t> = 0, \qquad <\zeta_t\zeta_s> = \delta(t-s) .$$ (6)

With Stratonovich calculus, a Fokker-Planck equation [1] can be set up for $P(x,t)$,

$$\frac{\partial}{\partial\tau}P = \frac{\partial}{\partial x}\left[\frac{\alpha}{2}\frac{\partial}{\partial x}(g^2P) - (g + \frac{\alpha}{2}gg')P\right] ,$$ (7)

where $\alpha = D^2/W$ is a noise parameter, and g is a function defined by,

$$g(x) = x\, 1 - (x/\Omega)^{n-1} .$$ (8)

We see that noise factor will always help to drift the peak and diffuse the width of $P(x,t)$. Moment equations can then be derived and expressed as,

$$\frac{d}{d\tau}\overline{x^k} = k<x^{k-1}g> + \frac{\alpha}{2}\left[<x^{k-1}gg'> + (k-1)<x^{k-2}g^2>\right] .$$ (9)

They can be truncated by using Eq.(4) to a closed form for the first two moments.

Equations similar to Eq.(5) can be derived, but are too lengthy to be displayed.

Stochastic behaviors are found to be similar to those of intrinsic case. Fluctuation in concentration is insignificant unless the noise parameter α is considerably large. In general, noise weakens stability and increases fluctuation in $\bar{x}(t)$. It is found that the survival SS with $x_s = \Omega$ will turn to be unstable, if α is larger than a critical value which is model dependent [4],

$$(\alpha_c)_{n=2} = 0.75 , \qquad (\alpha_c)_{n=3} = 0.5 . \tag{10}$$

Fluctuation catastrophe occurs before the system has a chance to relax to the SS. The result is that the expected SS replication is detoured to an extinction, even if the system is supposed to have an average rate $W > 0$.

For $\alpha = \alpha_c$, the SS is marginally stable. Within a narrow region of $\alpha < \alpha_c$, noises affect transient processes in a very unique way. The relaxation time diverges with an exponent which is equal to a mean field value of unity [4]. This is demonstrated in Fig.1 together with a plot of size-effect. The minimum population required to escape fluctuation catastrophe increases accordingly, i.e. the domain of attraction to the SS is Shrinking as $\alpha \rightarrow \alpha_c$.

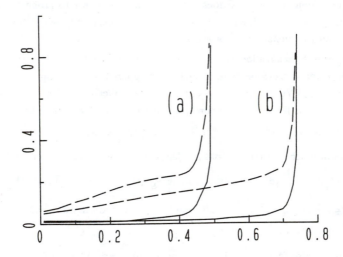

Fig. 1 Relaxaton time ($\tau_s/10^3$ in solid curves), and minimum concentration to have a secure relaxation (ξ_o in dashed curves) are plotted as function of noise parameter α. Both (a) cubic and (b) quadratic models are considered.

4. NONLINEAR RELAXATION TIME

Various types of characteristic times have been defined to characterize transient processes in nonequilibrium systems. Most of them bears stochastic meanings, for example, mean first pasage time [1] in multistable systems and switching time distribution [5] in laser systems.

The relaxation time investigated in last two sections is an operational one which describes the time period required practically to attain a SS. It applies to deterministic as well as stochastic treatments. A formulation of *nonlinear relaxation time* (NLRT) which can be defined in terms of any function related to transient, applies also for both treatments [6,7]. NLRT defined in terms of the moments can be expressed as,

$$T^{(k)} = \int_0^\infty dt \left[\overline{x^k}(t) - \overline{x_s^k} \right] \Big/ \left[\overline{x^k}(0) - \overline{x_s^k} \right]. \tag{11}$$

Numerically, NLRT can be integrated as a sum, with moments evaluated from moment equations for each time interval,

$$T^{(k)} = \sum_j \left[\Omega^k - \overline{x^k}(j\delta t) \; \delta t \right] \Big/ (\Omega^k - x_o^k), \tag{12}$$

where δt is the small time increment chosen for numerical integration. In the above, summation terminates so far as the SS criterion is satisfied.

The deterministic NLRT can also be defined from Eq.(11), by dropping out the over-head bars which represent averages. Results for the quadratic model can be integrated easily and are given by,

$$\tau^{(1)} = -(1 - \xi_o) \ln(\xi_o) \tag{13}$$

$$\tau(2) = (1 - \xi_o^2 - \ln\xi_o) / (1 - \xi_o), \tag{14}$$

where $\tau^{(k)} = WT^{(k)}$ is the deterministic NLRT.

Fig.2 shows the results of $T^{(1)}$ and $T^{(2)}$ for the quadratic model. As noise parameter α increases, $T^{(k)}$ also increase slowly and steadily. In the critical region NLRT bend down upruptly, instead of diverge as relaxation time does. It is due to the result that, as $\alpha \to \alpha_c$, $\overline{x}(t)$ approache $\overline{x}_s$ after passing through a maximum which is slightly larger than $\overline{x}_s = \Omega$. The critical slowing down should results in a divergence in NLRT if $\overline{x}(t)$ is a monotonic function of time.

The way $\overline{x}(t)$ approaches SS is probably a result of approximations employed during different stages of analysis. We stress here that Critical slowing down is a reasonable phenomenon, as would be demonstrated by a linear stability analysis of a SS which is drifting toward a marginal stability.

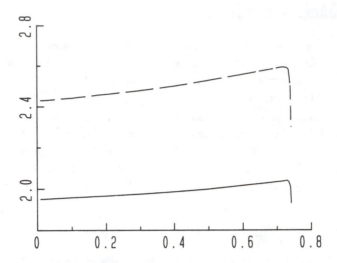

Fig. 2. NLRT defined in terms of the first (solid curve) and the second (dashed curve) moments are plotted against noise parameter α. Quadratic model is assumed.

The magnitude of NLRT itself, unlike the relaxation time, has no direct physical meaning other than an averaged properties of the defining function. It might loss many of the transient characteristics, especially when its defining function doesn't vary with time monotonically.

5. DISCUSSIONS AND CONCLUSIONS

In modeling a complex system, it is desirable to have stochastic analysis in addition to deterministic treatment, so that we could have an estimation on the effects of intrinsic and extrinsic fluctuations.

Nonequilibrium transient processes are very difficult to analyze. Moment expansion scheme enables us to derive the stochastic properties in an approximate way. With the truncated moment equations, we manage to probe the transient behaviors preceding a noise-induced instability and to evaluate NLRT numerically. This approximation remains valid so far as relative fluctuations are not significant.

It is interesting to find that noise effect depend quite sensitively on the non-linearity of model system. The fact that α_c is smaller for the n= 3 model implies that this cubic model is more vulnerable to noise effect. Furthermore, cubic model requires larger x_0 to avoid fluctuation catastrophe. In general cubic model is found to have larger fluctuation during transient.

Noises are unavoidable for open systems. However it can also be ultilized as a tool to probe noise induced phenomena in nonequilibrium systems. Critical slowing down, on which there are still conflicting views [7,8], would be an interesting topic. Much more effects are needed to investigate the stochastic transients, especially those near the nonequilibrium transitions.

This work is supported in part by the National Science Council of ROC through the contract number NSC79-0208-M008-39, and a travelling grant number 28111F. The hospitality of Professor Kitja Lindenberg is also gratefully acknowledged.

REFERENCES

1. M. Eigen, Naturwissenschaften **58**, 465 (1971).

2. C.W. Gardiner, Handbook of Stochastic Methods, 2nd ed. (Springer, Berlin, 1985).

3. H.K. Leung, Bull. Math. Biol. **42**, 231 (1985).

4. H.K. Leung, Phys. Rev. A**37**, 1341 (1988); A**41**, 1862 (1990).

5. G. Broggi, L.A. Lugiato, and A. Colombo, Phys. Rev. A**29**, 2949 (1984).

6. K. Binder, Phys. Rev. B**8**, 3423 (1973).

7. P.J. Jackson, C.J. Lambert, R. Mannella, P. Martano, P.V.E. McClintock, and N.G. Stock, Phys. Rev. A**40**, 2875 (1989).

8. L. Borci, L. Fronzoni, P. Grigolini, and V. Palleschi, Phys. Rev. A**39**, 2097 (1989).

Active Walker Models for Filamentary Growth Patterns

R.D. Freimuth and L. Lam

Many patterns in Nature or laboratory systems are made up of filaments, e.g. diffusion-liminted aggregates (DLA), electrodeposits, dielectric breakdowns, liquid crystal threads, and polymers. These are often modeled by the random walks as in the DLA model or its generalizations. A new approach is introduced here. Specifically, we consider an *active* walker capable of changing the landscape when walking on it, which is time dependent due to the (internal) influence of the walker, or external disturbance or noise; the movement of the walker is in turn influenced by the state of the landscape. In other words, our model consists of two components, the walker and the landscape, coupling to each other. (For a walk in space of three or higher spatial dimension, the word "environment", in place of "landscape", is more suitable. Since the computer results below involve only walks on a two-dimensional lattice, we shall use the word "landscape" throughout this paper.)

The case of a single walker on an initially flat surface is first considered; the walker's statistics are calculated and compared to those of a random walker. Generalizations to branching, biased walks, inhomogeneous background and multiwalkers are also included. Excellent qualitative agreement between the patterns from our computer model and the experimental dielectric breakdown of liquids in Hele-Shaw cells is obtained.

Finally, the relationships of our active walker model (AWM) with self-organized criticality in terms of the sandpile model and the surface-growth model are discussed.

R.D. Freimuth (1,2) and L. Lam (1): (1) Nonlinear Physics Group, Department of Physics, San Jose State University, San Jose, CA 95192-0106. (2) Present address: Department of Physics, University of California at San Diego, La Jolla, CA 92093.

1. INTRODUCTION

Generally speaking, growth patterns in Nature or laboratory systems [1,2] are either compact or filamentary. The former include viscous fingers in simple liquids [3] or liquid crystals [4,5], interfacial growths in directional solidification [6] or liquidcrystallization [7], flame propagation [8], and electrodeposits [9]. The latter include DLAs [10,11], electrodeposits [9,12], dielectric breakdowns [13-15], "nematoids" and smectic filaments [16], and polymer conformations [17].

Random walk has been used extensively [9-11,17,18] to model many of these filamentary patterns. In these efforts, the space in which the random walker walks is always passive and time independent, which could be either a fractal [19] or nonfractal landscape. This is in contrary to the situation in many real systems in which the filaments interact actively with the environment. Examples include the motion of the electrolyte and the small movement of the deposited filaments observed in electrodeposit formation in an open cell or a thick cell [9,20]; the "burning" of the chemicals on the inner surfaces of the glass plates and diffusion of heat when "burned" filaments are created in a Hele-Shaw cell of liquids [15]; the motion of the solution surrounding a polymer molecule.

To simulate these cases of interactive filament-enviroment systems, an active walker model is constructed. The model can be deterministic or probabilistic, depending on how the rule of selecting the next step for the walker is constructed (Sec. 2). In any case, it is *not* a random walker. Generalizations are presented in Sec. 3. In this paper, only the case of self-avoiding active walkers will be considered. The simpler case of self-crossing active walkers can be handled similarly.

Note that our AWMs are closely related to the models of self-organized criticality (Sec. 4).

2. THE BASIC MODEL: A SINGLE ACTIVE WALKER

Consider a lattice with a scalar field, called the "potential" $V(r_i,n)$, defined at each site i (with position r_i) at discrete time n (with n = 0,1,2,...). V is the sum of two parts: a possibly inhomogeneous, time-dependent background V_0 and an ever-updating part V_1 due to the influence of the active walker. Specifically, we assume that $V(r_i,n) = V_0(r_i,n) + V_1(r_i,n)$, with $V_1(r_i,0) = 0$ and

$$V_1(r_i,n+1) = V_1(r_i,n) + W(r_i\text{-}R(n)) \tag{1}$$

where W is a time-independent potential which may be aniotropic or isotropic, in the form given in Fig. 1; $R(n)$ is the position of the single walker at time n. The

304

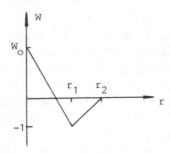

Fig. 1. An isotropic W(r) function. r is the magnitude of **r**.

next step of the walker is into one of the not-yet-visited neighboring sites $\mathbf{r}_j$ according to the potential differences between the presently occupied site and the neighboring sites, $V(\mathbf{R}(n),n+1) - V(\mathbf{r}_j,n+1) \equiv D_n(j)$. [Note that $\mathbf{R}(n) - \mathbf{r}_j$ is equal to a lattice constant vector if the site j is a nearest neighbor.] If $D_n(j) < 0$, then the walker stops at $\mathbf{R}(n)$ and the walk is terminated. For $D_n(j) \geq 0$, (i) in the *deterministic* active walker (DAW) model, $\mathbf{R}(n+1) = \mathbf{r}_j$ if $D_n(j)$ is the largest among all the possible neighbors; or, (ii) in the *probabilistic* active walker (PAW) model, the probability of $\mathbf{R}(n+1) = \mathbf{r}_j$ (i.e., site j being occupied) is proportional to $D_n(j)$. In the DAW model, if more than one neighboring sites have the same lowest D_n, then one of these sites is chosen randomly to be the $\mathbf{R}(n+1)$. Mathematically, these selection rules can be expressed as

$$\mathbf{R}(n+1) = f[D_n(j)] \qquad (2)$$

where f is a suitably defined operator.

A metaphor corresponding to Eq. (1) is a walker on a desert with initial landscape V_0; before the walker takes the next step, he rearranges the landscape around himself by redistributing some sand dictated by a fixed scheme W. (For W given in Fig. 1, he digs a circular trench around himself and dumps the sand over his head -- a *dumb* walker!) Then, he surveys this new landscape around him and chooses to step into the neighboring spot with the smallest height (the DAW model) or, being slightly drunk, he steps into one of the neighboring spots with a probability proportional to the depth of that spot as seen from his vantage point (the PAW model). In both cases, the walker only steps down but not up.

In some physical systems, there may be a delay for the effect of the walker on the landscape to manifest itself. This is the case of the *smart* walker who rearranges the landscape behind him (to avoid dumping sand on his head). In this case, Eq. (1) is generalized to

$$V_1(\mathbf{r}_i,n+1) = V_1(\mathbf{r}_i,n) + W(\mathbf{r}_i\text{-}\mathbf{R}(n\text{-}m)) \tag{3}$$

where m (= 0,1,2,...) is the time lag, and W = 0 if n < m. Equation (1) corresponds to m = 0.

While the AWMs proposed here are obviously applicable to many systems, to be specific and to gain more insights into the physical movtivations behind the models, let us use the dielectric breakdown of liquids in Hele-Shaw cells [15] as an example . In the experiments of Ref. [15], a conducting layer of indium tin oxide was coated on each of the inner surfaces of the two glass plates forming the Hele-Shaw cell. Oil or nematic liquid crystal as an example of simple or complex liquid, respectively, was put in the cell. When the voltage applied between the two plates is larger than a critical value dielectric breakdown of the liquid occurs and some dark "burned" tracks are left on the two inner surfaces. (See Ref. [15] for more details.) Note that a uniform electric field across the plates is maintained during the whole process.

The physics of this burning process is complicated and not entirely clear; it involves the nonlinear effects in macroscopic kinetics [21] and is closedly related to combustion phenomenon. Essentially, two intertwined processes -- dielectric breakdown and chemical combustion -- are involved. Above the voltage threshold, the liquid molecules are ionized due to dielectric breakdown. The resulting electrons and positive ions are attracted towards the two plates with positive and negative potentials, respectively. Bubbles due to the evaporation of the liquid is formed (as observed experimentally [15]). The electrons and ions, in the presence of heat, interact chemically with the material of the conducting layers resulting in the formation of the burned tracks. The chain reaction is sustained by heat diffusion, as well as by the movement and new creation of the charge carriers due to further breakdown. (The fact that the widths of the burned tracks left on the two plates differ from each other [15] implies that this is not a purely thermal process.)

With the W of Fig. 1, the central peak in W increases the potential at sites along the track making them unlikely to be visited again, mimicking the fact that the burned tracks cannot be burned again. The dip at r_1 lowering the potential in front of the track, together with the reinforcement due to the dips from the previous few steps, provides the tendency for the track to go forward -- the *forward-bias* effect. (In fact, if the peak maximum W_0 is large enough one obtains a self-avoiding walk without the self-avoiding process explicitly built into the program.) Simultaneously these dips dig up two trenches along the track so that once the track turns around, it has the tendency to grow parallel to the previous tracks -- the *track-affinity* effect due to the heating of the regions near the track,

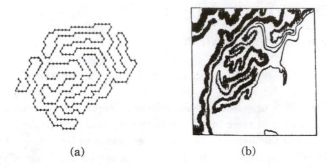

(a)   (b)

Fig. 2. (a) Computer result from the DAW model with $V_0 = 0$ and $m = 0$. (b) Experimental pattern obtained from a local blow-up of Fig. 8.

assuming that heat increases the chance of a chemical burning reaction. [These two effects are retained if a more realistic aniotropic W with the dip behind the walker eliminated is used, but the use of an isotropic W is more convenient. In fact, in the rest of this paper, the isotropic W of Fig. 1 will be used; also, simple square lattice will be assumed, and only nearest-neighbor (nn) empty sites of $\mathbf{R}(n)$ will be allowed as possible sites for $\mathbf{R}(n+1)$.]

Fig. 2 (a) shows the computer result from the DAW model with no time lag ($m = 0$), which closely resembles the experimental pattern of Fig. 2 (b) where the track-affinity effect is clearly shown.

Some results from the PAW model are presented in Figs. 3 and 4. In general, the walk length increases with W_0; small ρ ($\equiv r_1/r_2$) gives more curly, tightly packed walks, while large ρ gives longer, strighter walks; medium value of r_2 (~ 15) gives curly, short walks, and large r_2 (~ 40) gives wider, longer walks. It should be pointed out that the effects of all these parameters are coupled to each other. Also, for the same set of parameters quite different forms of walks can result (see Fig. 4).

Fig. 3. A spiral walk obtained from the PAW model. $V_0 = 0$; $m = 1$, $W_0 = 5$, $r_2 = 15$ and $\rho = 0.3$.

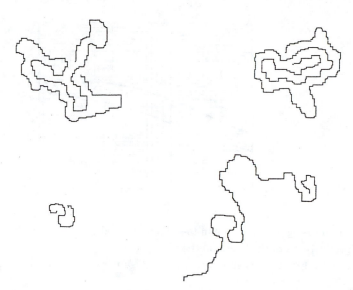

Fig. 4. Four different walks obtained from the PAW model. $V_0 = 0$; $m = 1$, $W_0 = 10$, $r_2 = 25$ and $\rho = 0.3$. (In Figs. 7 and 9, $m = 0$.)

The statistics of 100 walks from the PAW model with another set of parameters are shown in Fig. 5; the walks are short and compact. The log-log plots of the R_e and R_g vs time n are given in Fig. 6, where R_e and R_g are, respectively, the ensemble-averaged end-to-end length and the radius of gyration of each walk (before or until it naturally terminates). We find that $R_g \sim n^v$ (and $R_e \sim n^v$ approximately) with $v = 0.50$ which is the same as that for a random walk. This value of v is due to the relatively small value of ρ used for which $r_1 = 1.25$; it shifts to 0.81 when ρ becomes 0.3 with the rest of the parameters unchanged.

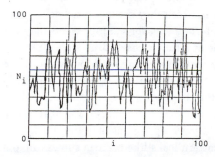

Fig. 5. Maximum length of the walk N_i for the i-th run from the PAW model, with the same parameters as in Fig. 4 except $\rho = 0.05$. Averaged N_i is 50.15.

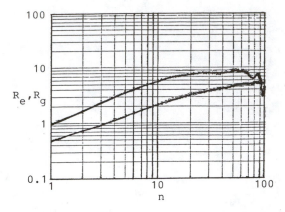

Fig. 6. Log-log plots of the end-to-end length and radius of gyration, R_e and R_g respectively, vs n, the length of the walk. The upper curve is R_e. The walks are those from Fig. 5. Size effect appears at large n; exponent in the text is obtained from the straight-line part.

3. GENERALIZATIONS

The basic model above was generalized by LAM et al [22] in several directions. First, the single walker is allowed to branch. For example, after the $\mathbf{R}(n+1)$ $(= \mathbf{r}_k$, say) site is chosen as specified in Sec. 2, other nn sites $\mathbf{r}_j$ of $\mathbf{R}(n)$ with $D_n(j) \geq 0$ and $D_n(j) \geq \gamma D_n(k)$ are also allowed to be occupied, where γ is the branching factor, a prescribed constant parameter; each newly occupied new site immediately after $\mathbf{R}(n)$ is considered as a branched active walker and can be treated independently. These branched walkers in fact influence the movement of each other through their restructuring interaction with the landscape. If these branched walkers are activated in sequence, as is assumed in this paper, it is obvious that the order in which they are activated does influence the morphology of tracks formed. (Alternatively, for example, one may activate the branched walkers randomly, one at a time at each time step.)

In Fig. 7(a), a track pattern starting with four walkers forming a solid square at the center of a 100x100 square lattice is shown. The background potential V_0 is inhomogeneous, in the shape of an upward cone (i.e., with the sharp end of the cone pointing upward) with maximum energy 100 at the center of the lattice; the cone extends downward touching the edges of the the square lattice with energy zero. The resulting pattern looks like a Paris subway map and is DLA-like; the fractal dimension D obtained from $N_r \sim r^D$ [Fig. 7(b)] is D = 1.69, where N_r is the number of points within a circle of radius r centering at the lattice center. Note

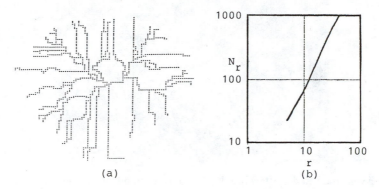

Fig. 7. DLA-like pattern generated by the PAW model with branching. Square lattice of 100x100 size is used. V_0 is a fixed upward cone (see text). $W_0 = 20$, r_2 = 10, $\rho = 1.0$; $\gamma = 0.8$. (a) The DLA-like morphology. (b) The log-log plot of N_r vs r. Note the slight crossover at r = 10.

that this value of D is consistent with that of the DLA, D = 1.70 [10,11]. More statistics are needed to determine whether this result is purely accidental.

When the upward cone in Fig. 7(a) is replaced by an downward one, it is possible to obtain [22] spiral tracks similar to those observed by MOGI et al [23] in the growth of silver metal leaves in a high magnetic field. If, instead, the background V_0 is an inclined plane, one obtains a walk with bias in one direction [22].

Second, multiple walkers are used. In the experiment of Ref. [15], due to the presence of the uniform field across the plates, dielectric breakdown may occur at more than one spot. In order to simulate such a pattern observed in the dielectric breakdown of liquid in a Hele-Shaw cell (Fig. 8), seven walkers placed randomly in a 100x100 square lattice (at time n = 0) are allowed to walk with branching using the PAW model. A typical pattern so obtained is shown in Fig. 9. Although the exact pattern (with the same initial positions of the seven walkers) differs in each run, the overall features are the same and compare very well with that in Fig. 8. These overall features are: (i) There are but not too much branching. (ii) Both the forward-bias and track-affinity effects are there. (iii) There are some occasionally terminated tracks. (iv) There exist large enclosed and open spaces. The qualitative agreement between the computer and experimental results here indicates that the AWM does capture the essential physics of the dielectric breakdown processes. More quantitative results are needed.

Other generalizations are possible. For example, the form of the W function can be drastically changed and may be time dependent (see Sec. 4); upward climbing of the walker may be allowed so that the motion of the walker is closer

Fig. 8. Experimental dielectric breakdown pattern of oil in a Hele-Shaw cell (Ref. [15]). The scale shown is in mm.

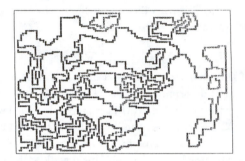

Fig. 9. A multiwalkers track pattern with branching from the PAW model. Seven walkers are placed randomly initially. $V_0 = 0$; $W_0 = 15$, $r_2 = 20$, $\rho = 0.3$; $\gamma = 0.95$.

to that of a ball rolling in a sanddune or a sandpile. In the latter case with the DAW, the model can be described by an integral equation [20]. Temperature effects can be incorporated by replacing the probability of selecting neighboring sites which is proportional to the energy difference, by the use of the Boltzman factor. A time dependent V_0 may be used to mimic effects of noise or wind blowing on the landscape.

4. RELATIONSHIPS WITH SELF-ORGANIZED CRITICALITY

The sandpile model of self-organized criticality (SOC) of BAK, TANG and WIESENBFELD [24,25] may be considered as an AWM. The sandpile is our landscape; the adding and toppling of sand are the actions due to the walker. For

311

example, for the 1D model, one puts a walker randomly on a sandpile described by $V_0(\mathbf{r})$; the $W(r,n)$ function is now discrete and time dependent, given by $W(0,0) = 1$ and $W(-1,0) = -1$, if $V(\mathbf{R}(n),0) \leq V_c$; $W(0,0) = -1$, $W(-1,0) = 0$ and $W(1,0) = 1$, if $V(\mathbf{R}(n),0) > V_c$; in both cases, $W(r,0) = 0$ otherwise. At any time n, branching to a nn site is mandatory if the V at that site is disturbed. For $n > 0$, $W(r,n) = 0$ if $V(\mathbf{R}(n),n) \leq V_c$; $W(0,n) = -2$ and $W(\pm1) = 1$, if $V(\mathbf{R}(n),n) > V_c$. A branched walker remains stationary when the V's of all the nn sites are not disturbed; it may be reactivated again if the situation changes in the future. (Here, V stands for z of Ref. [24].) Note that the $W(r,0)$ function combines the action of adding a grain of sand and the possible subsequent toppling at the site where the sand is added. The $W(r,n)$ function, $n > 0$, allows the branched walker to check whether a site needs to be toppled or not. The total length of all the walkers represents the number of sites disturbed (which is usually larger than the number of toppling sites) since adding the single grain of sand. The SOC state is achieved by repeating the above process many times with the previous walker removed. Generalization to the 2D case is straightforward.

The surface-growth model of SANDER and YAN [26] is also an AWM. The rule of changing the landscape is different from those of the sandpile model, but a SOC state is still achieved. Recasted in the language of our AWM, the relandscaping corresponds to using a constant $W(r)$ with $W(r) = 0$ if $V(r,n) \neq d$; $W(0) = -2d$, $W(\pm1) = 1$, and $W(r) = 0$ otherwise if $V(r,n) = d$, where d is the dimension of the lattice. (Here, V stands for n of Ref. [26].) The walker starts randomly and is terminated immediately after it rearranges the landscape, and then removed (i.e., it does not walk, and there is no avalanche or dissipation). The process is repeated many times. Motivated by these results, we believe that if the walker in the basic AWM of Sec. 2 is added repeatedly and randomly, where walking is allowed, one can also obtain a SOC state in the landscape. The same goes if multiwalkers with or without branching are used. Work along this line is in progress.

Relationships between our AWMs and other complex system models, such as the rugged landscape model used in evolutionary biology, computer science and spin glass [27] will be discussed elsewhere.

This work is supported by the NSF-REU program and a Syntax Corporation grant from the Research Corporation. We thank J.T. Fredrick, D.A. Kayer, R.D. Pochy and M.K. Pon for helpful discussions and assistance in producing some of the diagrams.

REFERENCES

1. *Nonlinear Structures in Physical Systems: Pattern Formation, Chaos and Waves*, edited by L. Lam and H.C. Morris (Springer, New York, 1990).
2. *Random Fluctuations and Pattern Growth: Experiments and Models*, edited by H.E. Stanley and N. Ostrowsky (Kluwer, Boston, 1988).
3. P.G. Saffman, in Ref. [1].
4. L. Lam, in *Wave Phenomena*, edited by L. Lam and H.C. Morris (Springer, New York, 1989).
5. L. Lam et al, Liq. Cryst. **5**, 1813 (1989).
6. E.A. Brener and V.I. Mel'nikov, Adv. Phys. **40**, 53 (1991); D. Kessler, J. Koplik and H. Levine, Adv. Phys. **37**, 255 (1988); J.S. Langer, in *Chance and Matter*, edited by J. Scouletie, J. Vannimenus and R. Stora (North-Holland, Amsterdam, 1987).
7. J.-M. Flesselles, A.J. Simon and A.J. Libchaber, Adv. Phys. **40**, 1 (1991); P. Oswald, J. bechhoefer and F. Melo, MRS Bulletin **56**, 38 (1991); L. Lam, in *Nonlinear and Chaotic Phenomena*, edited by W. Rozmus and J.A. Tuszynski (World Scientific, Teaneck, 1991).
8. *Dynamics of Curved Fronts*, edited by P. Pelcé (Academic, San Diego, 1988).
9. L. Lam, R.D. Pochy and V.M. Castillo, in Ref. [1].
10. T. Vicsek, *Fractal Growth Phenomena* (World Scientific, Teaneck, 1989).
11. T.A. Witten and L.M. Sander, Phys. Rev. Lett. **47**, 1400 (1981); L.M. Sander, in *Introduction to Nonlinear Physics*, edited by L. Lam (Springer, New York, 1992).
12. L.M. Sander, in *The Physics of Structure Formation*, edited by W. Güttinger and G. Dangelmayr (Springer, New York, 1987).
13. A.R. von Hippel, *Dielectric and Waves* (Wiley, New York, 1954).
14. I. Adamczewski, *Ionization, Conductivity and Breakdown in Dielectric Liquids* (Taylor and Francis, London, 1969).
15. L. Lam, R.D. Freimuth and H.S. Lakkaraju, Mol. Cryst. Liq. Cryst. **199**, 249 (1991).
16. A. Adamczyk, Mol. Cryst. Liq. Cryst. **170**, 53 (1989); S.L. Arora, P. Palffy-Muhoray, R.A. Vora, D.J. David and A.M. Dasgupta, Liq. Cryst. **5**, 133 (1989).
17. M. Doi and S.F. Edwards, *The Theory of Polymer Dynamics* (Clarendon, Oxford, 1986).
18. V.M. Castillo, R.D. Pochy and L. Lam, in *Applications of Statistical and Field Theory Methods to Condensed Matter*, edited by D. Baeriswyl, A.R. Bishop and J. Carmelo (Plenum, New York, 1990); R.D. Pochy, A. Garcia, R.D. Freimuth, V.M. Castillo and L. Lam, Physica **51**, 539 (1991).
19. P. Meakin, Phys. Rev. A **36**, 2833 (1987); J. Phys. A **20**, L771 (1987).

20. L. Lam, unpublished.
21. A.G. Merzhanov and E.N. Rumaov, Sov. Phys. Usp. **30**, 293 (1987).
22. L. Lam, R.D. Freimuth, M.K. Pon, D.A. Kayser, J.T. Fredrick and R.D. Pochy, in *Physics of Pattern Formation in Complex Dissipative Systems*, edited by S. Kai (World Scientific, River Edge, 1992).
23. I. Mogi, S. Okubo and Y. Nagakawa, J. Phys. Soc. Jpn. **60**, 3200 (1991).
24. P. Bak, C. Tang and K. Wiesenfeld, Phys. Rev. Lett. **59**, 381 (1987); Phys. Rev. A **38**, 364 (1988).
25. P. Bak, Comput. Phys. **5**, 430 (1991).
26. L.M. Sander and H. Yan, Phys. Rev. A **44**, 4885 (1991).
27. S. Kauffman, in *Lectures in the Sciences of Complexity*, edited by D.L. Stein (Addison-Wesley, Reading, 1989); E.D. Weinberger, Phys. Rev. A **44**, 6399 (1991), and references therein.

Index of Contributors